NIOSH HAZARD REVIEW

Occupational Hazards in Home Healthcare

DEPARTMENT OF HEALTH AND HUMAN SERVICES
Centers for Disease Control and Prevention
National Institute for Occupational Safety and Health

This document is in the public domain and may be freely copied or reprinted.

Disclaimer

Mention of any company or product does not constitute endorsement by the National Institute for Occupational Safety and Health (NIOSH). In addition, citations to Web sites external to NIOSH do not constitute NIOSH endorsement of the sponsoring organizations or their programs or products. Furthermore, NIOSH is not responsible for the content of these Web sites.

Ordering Information

To receive documents or other information about occupational safety and health topics, contact NIOSH at

>Telephone: **1–800–CDC–INFO** (1–800–232–4636)
>TTY: 1–888–232–6348
>E-mail: cdcinfo@cdc.gov

>or visit the NIOSH Web site at **www.cdc.gov/niosh**.

>For a monthly update on news at NIOSH, subscribe to NIOSH *eNews* by visiting **www.cdc.gov/niosh/eNews**.

>DHHS (NIOSH) Publication No. 2010–125

January 2010

SAFER • HEALTHIER • PEOPLE™

Foreword

An aging population and rising hospital costs have created new and increasing demand for innovative healthcare delivery systems in the United States. Home healthcare provides vital medical assistance to ill, elderly, convalescent, or disabled persons who live in their own homes instead of a healthcare facility, and is one of the most rapidly expanding industries in this country. The Bureau of Labor Statistics projects that home healthcare employment will grow 55% between 2006–2016, making it the fastest growing occupation of the next decade.

Home healthcare workers facilitate the rapid and smooth transition of patients from a hospital to a home setting. They offer patients the unique opportunity to receive quality medical care in the comfort of their own homes rather than in a healthcare or nursing facility.

Home healthcare workers, while contributing greatly to the well-being of others, face unique risks on the job to their own personal safety and health. During 2007 alone, 27,400 recorded injuries occurred among more than 896,800 home healthcare workers.

Home healthcare workers are frequently exposed to a variety of potentially serious or even life-threatening hazards. These dangers include overexertion; stress; guns and other weapons; illegal drugs; verbal abuse and other forms of violence in the home or community; bloodborne pathogens; needlesticks; latex sensitivity; temperature extremes; unhygienic conditions, including lack of water, unclean or hostile animals, and animal waste. Long commutes from worksite to worksite also expose the home healthcare worker to transportation-related risks.

This document aims to raise awareness and increase understanding of the safety and health risks involved in home healthcare and suggests prevention strategies to reduce the number of injuries, illnesses, and fatalities that too frequently occur among workers in this industry.

John Howard, M.D.
Director, National Institute for
  Occupational Safety and Health
Centers for Disease Control and Prevention

Contents

Foreword iii
Abbreviations ix
Acknowledgments x

Chapter 1 | Background 1
    1.1   References 2

Chapter 2 | Musculoskeletal Disorders and Ergonomic Interventions 3
    2.1   Introduction 3
    2.2   What is the impact of musculoskeletal disorders on the home healthcare industry? 3
    2.3   What are the risk factors for developing musculoskeletal disorders? . . . 3
        2.3.1   What are some factors that complicate patient transfers? 4
        2.3.2   What factors contribute to awkward postures? 4
        2.3.3   What other factors contribute to musculoskeletal disorders? . . . 5
    2.4   Can anything help limit musculoskeletal disorders? 5
    2.5   What can I do to prevent musculoskeletal disorders? 6
        2.5.1   Recommendations for Employers 7
        2.5.2   Recommendations for Workers 10
    2.6   Resources 10
    2.7   References 11

Chapter 3 | Latex Allergy 15
    3.1   Introduction 15
    3.2   What are some sources of exposure to latex? 15
    3.3   What are the effects of latex exposure? 15
        3.3.1   What is irritant contact dermatitis? 15
        3.3.2   What is allergic contact dermatitis? 16
        3.3.3   What is latex allergy? 16
    3.4   What are some products that contain latex? 17
    3.5   How can I prevent exposure to latex? 17
        3.5.1   Recommendations for Employers 17
        3.5.2   Recommendations for Workers 18

3.6	Resources	18
3.7	References	19

Chapter 4 | Exposure to Bloodborne Pathogens and Needlestick Injuries 21

4.1	Introduction	21
4.2	How serious is the risk of exposure from needlestick and sharps injuries?	21
4.3	What regulations should I be aware of?	22
4.4	What about needleless systems and needle devices with safety features?	23
	4.4.1 What needleless systems and needle devices with safety features are available?	23
	4.4.2 How do I select and evaluate needleless systems and needle devices with safety features?	25
4.5	What can I do to prevent and control needlestick and sharps injuries?	25
	4.5.1 Recommendations for Employers	25
	4.5.2 Recommendations for Workers	26
4.6	What should I do if I am exposed to the blood of a patient?	26
4.7	Resources	27
4.8	References	27

Chapter 5 | Occupational Stress 29

5.1	Introduction	29
5.2	What are some specific stressors of home healthcare workers?	29
5.3	What can I do to prevent and control occupational stress?	30
	5.3.1 Recommendations for Employers	30
	5.3.2 Recommendations for Workers	30
5.4	Resources	31
5.5	References	31

Chapter 6 | Violence 33

6.1	Introduction	33
6.2	What are some factors that increase the risk of violence to home healthcare workers?	33
6.3	What does workplace violence include?	33

6.4	What are some effects of this violence?	34
6.5	How can I prevent and control violence in a patient's home?	34
	6.5.1 Recommendations for Employers	34
	6.5.2 Recommendations for Workers	35
6.6	Resources	37
6.7	References	37

Chapter 7 | Other Hazards ... 39

7.1	Introduction	39
7.2	What can I do to prevent and control the occurrence of or exposure to these hazards?	39
	7.2.1 Infectious Diseases	39
	7.2.1.1 Recommendations for Employers	39
	7.2.1.2 Recommendations for Workers	40
	7.2.2 Animals	40
	7.2.2.1 Recommendations for Employers	40
	7.2.2.2 Recommendations for Workers	40
	7.2.3 Home Temperature	40
	7.2.3.1 Recommendations for Employers	40
	7.2.3.2 Recommendations for Workers	40
	7.2.4 Hygiene	40
	7.2.4.1 Recommendations for Employers	40
	7.2.4.2 Recommendations for Workers	41
	7.2.5 Lack of Water	41
	7.2.5.1 Recommendations for Employers	41
	7.2.5.2 Recommendations for Workers	41
	7.2.6 Falls	41
	7.2.6.1 Recommendations for Employers	41
	7.2.6.2 Recommendations for Workers	41
	7.2.7 Severe Weather	41
	7.2.7.1 Recommendations for Employers	42
	7.2.7.2 Recommendations for Workers	42
	7.2.8 Chemical Spills and Acts of Terrorism	43
	7.2.8.1 Recommendations for Employers	43
	7.2.8.2 Recommendations for Workers	43

		7.2.9	Automobile Travel	44
			7.2.9.1 Recommendations for Employers	44
			7.2.9.2 Recommendations for Workers	44
	7.3	Resources		44
	7.4	References		44

Chapter 8 | Conclusions 47

    8.1   Checklists for Home Healthcare Workers' Safety. 48

Abbreviations

BLS	Bureau of Labor Statistics
CDC	Centers for Disease Control and Prevention
FDA	Food and Drug Administration
FEMA	Federal Emergency Management Agency
GPS	global positioning system
HBV	hepatitis B virus
HCV	hepatitis C virus
HIV	human immunodeficiency virus
IV	intravenous
NIOSH	National Institute for Occupational Safety and Health
NRL	natural rubber latex
OSHA	Occupational Safety and Health Administration
SOII	Survey of Occupational Injuries and Illnesses
TB	tuberculosis
VA	U.S. Department of Veterans Affairs

Acknowledgments

This document was prepared by the NIOSH Education and Information Division (EID), Paul Schulte, Ph.D., Director. Laura Hodson; Traci Galinsky, Ph.D.; Bonita Malit, M.D.; Henryka Nagy, Ph.D.; Kelley Parsons, Ph.D.; Naomi Swanson, Ph.D.; and Tom Waters, Ph.D. were the principle authors. The authors acknowledge Sherry Baron, M.D.; Barbara Dames; Sherry Fendinger; Christy Forrester; Michael Colligan, Ph.D.; James Collins, Ph.D.; Paula Grubb; Regina Pana-Cryan, Ph.D.; Robert Peters; Edward Petsonk, M.D.; and Joann Wess for contributing to the technical content of this document.

The authors thank Susan Afanuh, Vanessa Becks Williams, Elizabeth Fryer, and John Lechliter for their editorial support and contributions to the design and layout of this document.

Special appreciation is expressed to the following individuals and organizations for their external reviews and valuable comments:

Steven Christianson, D.O., M.M.
VNS HomeCare
New York, NY

Catherine Galligan, MS
University of Massachusetts
Lowell, MA

Lisa Gorski, MS, APRN, BC, CRNI, FAAN
Wheaton Franciscan
Home Health and Hospice
Mequon, WI

Elise M. Handelman, RN, M.Ed., FAAOHN
Occupational Safety and
  Health Administration
Washington, D.C.

Tina Marrelli, MSN, MA, RN
Editor Home HealthCare Nurse
The Journal for the Home Care
  and Hospice Professional
Boca Grande, FL

Kathleen M. McPhaul, PhD, MPH, RN
University of Maryland School
  of Nursing
Baltimore, M.D.

Doris Mosocco, RN, BSN, CHCE, COS-C
Heartland Home Health and Hospice
Williamsburg, VA

Rosemary K. Sokas, M.D., MOH
University of Illinois
School of Public Health
Chicago, IL

Wayne Young, B.A., M.B.A.
Service Employees International Union
Washington, D.C.

Background

Home healthcare workers help ill, elderly, convalescent, or disabled persons who live in their own homes instead of in a healthcare facility. Home healthcare workers encompass a variety of occupations, including nurses, home healthcare aides, physical therapists, occupational therapists, speech therapists, therapy aides, social workers, and hospice care workers. Under the direction of medical staff, they provide health-related services. The services may include helping with activities of daily living (for example, bathing, dressing, getting out of bed, and eating); delivering medical services such as administering oral, intravenous, or other parenteral medications; changing nonsterile dressings; giving massages or alcohol rubs; or helping with ventilators, braces, or artificial limbs. Home healthcare workers are predominantly female (89%) with 24.4% self identified as black or African American, 20.0% as Hispanic or Latino, and 4.4% as Asian [BLS 2008a]. Home healthcare workers may work any hour of the day or night and on any day of the week [NIOSH 1999; BLS 2008b].

Home healthcare is one of the most rapidly growing industries in the United States. According to the Bureau of Labor Statistics (BLS), 896,800 workers were employed in home healthcare services in 2007, and the number of workers is expected to grow by 55% between 2006–2016 [BLS 2008b]. The demand for home services is rapidly growing in this country for several reasons including: an increase in the aging population; hospitals providing more services on an outpatient basis; a decrease in the length of hospital stays; patients' preference for care in the home; and substantial cost savings to the health care system.

The rate of turnover is very high among healthcare workers, particularly home healthcare workers. Stonerock [1997] has reported turnover rates as high as 75% among home healthcare workers in some parts of the country and noted that within the labor pool from which home healthcare workers are drawn, other service occupations often compete more favorably. Attracting workers and retaining them is therefore a high priority for many home healthcare agencies, and providing a more healthful, less stressful, work climate is an important part of any retention strategy.

Some hazards that home healthcare workers may encounter are unique to the home setting. The work environment generally is not under the control of either the employer or the employee. Therefore, the home healthcare worker may encounter unexpected and unpredictable hazards, such as animals, loaded firearms or other weapons, and violence in the home, apartment building, or neighborhood. Persons other than the patient who are residing or visiting in the home may also be a risk to the worker.

Falls may occur when home healthcare workers are walking on ice- and snow-covered streets, driveways, sidewalks, and paths to the homes of their patients [BLS 1997].

Driving from home to home exposes the home healthcare worker to risks of vehicular injury or fatality.

According to BLS, there were 27,400 recordable injuries to home healthcare workers during 2007 resulting in an incidence rate of 4.3 per 1,000 full-time equivalent workers [BLS 2008c]. Sprains and strains were the most common lost-work-time injuries [BLS 2008d].

This document provides information about a number of potential hazards to home healthcare workers including muscloskeletal disorders, latex allergy, bloodborne pathogens, occupational stress, violence, and other work-related hazards. The document provides an overview of the hazards and provides recommendations for both employers and workers to eliminate the hazards or minimize risks. Understanding the challenges and implementing the suggested prevention strategies can reduce the number of injuries, illnesses and fatalities occuring among home healthcare workers.

1.1 References

BLS [1997]. Injuries to caregivers working in patients' homes. Issues in Labor Statistics, Summary 97–4. Washington, DC: U.S. Department of Labor, Bureau of Labor Statistics.

BLS [2008a]. Table 18. Employed persons by detailed industry, sex, race, and Hispanic or Latino ethnicity, 2007. Washington, DC: U.S. Department of Labor, Bureau of Labor Statistics [www.bls.gov/cps/cpsaat18.pdf].

BLS [2008b]. Career Guide to Industries, 2008-09 Edition, Health Care [www.bls.gov/oco/cg/cgs035/htm].

BLS [2008c]. Table 1. Incidence rates of nonfatal occupational injuries and illnesses by industry and case types, 2007. Washington, DC: U.S. Department of Labor, Bureau of Labor Statistics [www.bls.gov/iif/oshwc/osh/os/ostb1917.txt]

BLS [2008d]. Table R5. Incidence rates for nonfatal occupational injuries and illnesses involving days away from work per 10,000 full-time workers by industry and natures of injury or illness, 2007. Washington, DC: U.S. Department of Labor, Bureau of Labor Statistics [www.bls.gov/iif/oshwc/osh/case/ostb1947.txt].

NIOSH [1999]. The Answer Group. NIOSH: home healthcare workers. Written summary and videotapes of focus group meetings of home healthcare workers (June 13 and July 7, 1999) and Chicago, Illinois (June 28, 1999). Cincinnati, OH: U.S. Department of Health and Human Services, Centers for Disease Control, National Institute for Occupational Safety and Health.

Stonerock C [1997]. Home health aides: home care's "endangered species." Home Care Provid 2(1):15–17.

Musculoskeletal Disorders and Ergonomic Interventions

2.1 Introduction

All healthcare workers who lift and move patients are at high risk for back injury and other musculoskeletal disorders [Owen 1999; Waters et al. 2006]. A work-related musculoskeletal disorder is an injury of the muscles, tendons, ligaments, nerves, joints, cartilage, bones, or blood vessels in the extremities or back that is caused or aggravated by work tasks such as lifting, pushing, and pulling [Orr 1997]. Symptoms of musculoskeletal disorders include pain, stiffness, swelling, numbness, and tingling.

Home healthcare workers do many of the same tasks as workers in traditional healthcare settings, but conditions in the home setting often make the work more difficult. For instance, home healthcare workers most often perform heavy work, like lifting and moving patients, without assistance [Myers et al. 1993].

The following sections define the scope of the problem, discuss risk factors for developing musculoskeletal disorders in home healthcare work, and suggest ways to prevent musculoskeletal disorders.

2.2 What is the impact of musculoskeletal disorders on the home healthcare industry?

Work-related musculoskeletal disorders are a serious problem in the home healthcare industry [Galinsky et al. 2001]. Sprains and strains were the most common lost-work-time injuries to home healthcare workers in 2007 [BLS 2008a]. Home healthcare workers may injure themselves when transferring patients into and out of bed or when assisting patients walking or standing [El-Askari 1999]. The rate of injury from lifting in 2007 for home healthcare workers was 20.5 per 10,000 workers [BLS 2008b]. Compared with other workers, home healthcare workers take more frequent sick leave as a result of work-related musculoskeletal symptoms [Brulin et al. 1998a; Moens et al. 1994; Ono et al. 1995].

2.3 What are the risk factors for developing musculoskeletal disorders?

Healthcare workers can develop musculoskeletal disorders from any number of common work activities [NIOSH 1997], including the following:

- Forceful exertions (activities that require a person to apply high levels of force, such as during lifting, pushing, or pulling heavy loads)
- Awkward postures when lifting
- Repeated activities without adequate recovery time

Patient-handling tasks often involve motions that challenge a home healthcare worker's body including twisting, bending, stretching, reaching, and other awkward postures. The most frequent causes of back pain and other injuries among nursing staff (in home healthcare and in hospitals) are lifting and moving patients ("patient transfers") and bathing, dressing, and feeding patients [Orr 1997; NIOSH 1999; Owen 1999; Galinsky et al. 2001]. Healthcare workers who spend the most time transferring, bathing, and dressing patients have the highest rates of musculoskeletal injuries [Moens et al. 1994; Zelenka et al. 1996; Nelson et al. 1997]. In a NIOSH survey study of home healthcare workers, these tasks were identified as significant predictors of pain in the back, neck, shoulders, legs and feet, after adjusting for other factors such as the workers' age, weight, and physical activities outside of work [Waters et al. 2006]. Dellve et al. [2003] found that frequent heavy lifting, lifting in awkward postures, and lifting without assistance were significant predictors of permanent work disability in home healthcare workers.

2.3.1 What are some factors that complicate patient transfers?

- Incapacity is common among home healthcare patients; about 40% of them have one or more functional limitations because patients are being released after shorter hospital stays and require more intensive care during recovery at home [Jarrell 1997].

- Healthcare workers are commonly required to lift and move patients weighing 90 to 250 pounds. These weights exceed the NIOSH safe lifting limits for both men and women [Waters et al. 1993].

- The body weight of a patient is not evenly distributed, nor does a body have convenient hand-holds.

- The patient may be connected to a catheter, I.V., or other equipment, resulting in awkward postures for workers involved in his or her transfer.

- The functional limitations of the patient—physical, mental, or both—may interfere with the lift:
    — The patient may not be able to hold himself or herself up.
    — The patient may not be cooperative.
    — The patient may be obese (body mass index > 30) [Nelson et al. 2003].

- Certain lifting techniques used to minimize the load on the back may increase the load on other body parts such as the neck, shoulders, and arms [Knibbe and Friele 1996].

2.3.2 What factors contribute to awkward postures?

- Rooms in patients' homes are often small or crowded, and workers must often use awkward postures during patient care and transfer tasks [Myers et al. 1993]. Between 40 and 48% of the home healthcare workers' time may be spent in poor posture combinations, including forward-bent and twisted postures that are associated with shoulder, neck, and back complaints [Pohjonen et al. 1998; Torgen et al. 1995; Brulin et al. 1998b]. Shoulder and neck symptoms in home healthcare workers have been shown to be due to poor postures and forceful exertions during patient care tasks [NIOSH 2004; Elert et al. 1992; Johansson 1995; Torgen et al. 1995; Knibbe and Friele 1996; Brulin et al. 1998a; Meyer and Muntaner 1999].

- Beds may not be adjustable, preventing the worker from raising or lowering the patient to the best position for a proper lift. Owen [2003] found that problems with the bed's height, width, placement, and nonadjustability were frequently cited by home healthcare workers as major sources of back stress.

2.3.3 What other factors contribute to musculoskeletal disorders?

- Patients' homes usually do not have equipment to help with transfers.
- Home healthcare workers frequently endure long periods of standing or walking.

2.4 Can anything help limit musculoskeletal disorders?

The science of work design is called ergonomics. Ergonomics is the design of the work setting (including furniture, tools, equipment, and tasks) to help position the worker in a way that will lessen the possibility of injury when performing work tasks. Therefore, the ergonomics approach optimizes the worker's safety, health, and performance.

Researchers have found that help from a second trained person reduces the risk of injury during patient-handling tasks but not enough to make the tasks acceptably safe. Marras et al. [1999] concluded that manual patient handling is "an extremely hazardous job that had substantial risk of causing a low-back injury whether with one or two patient handlers." For this reason, ergonomic intervention, including the use of electronic and mechanical devices to help with patient transfers, is the most promising approach for reducing low-back injuries during patient handling.

Comprehensive ergonomic interventions using appropriate equipment and training have resulted in dramatic reductions in the incidence and severity of musculoskeletal injuries among healthcare workers. For example, in one study [NIOSH 1999], a "zero-lift" program was implemented in seven nursing homes and one hospital to eliminate manual patient transfers: Hoists and other equipment were used to lift patients rather than lifting manually. Injuries related to patient transfers were reduced 39%–79%. Other reductions were noted in the average number of lost workdays (86%), restricted workdays (64%), and workers' compensation costs (84%). In a review of patient-handling intervention research, Hignett [2003] identified 21 studies, conducted from 1982 through 2001, that evaluated patient-handling equipment and equipment training. Of the 21 studies, 16 (76%) reported positive effects including reductions in injuries, lost workdays, spinal loads, harmful postures, perceived exertion, and staffing requirements. Subsequent studies have cited similar positive effects for healthcare workers as well as positive effects on the quality of patient care [Ronald et al. 2002; Spiegel et al. 2002; Evanoff et al. 2003; Collins et al.

2004; Chhokar et al. 2005; Engst et al. 2005; Fujishiro et al. 2005; Santaguida et al. 2005; Nelson et al. 2006; Nelson et al. 2008]. Nelson et al. [2003] summarize numerous other case studies using ergonomic interventions in hospitals and nursing homes that have also shown large reductions in injury rates, workers' compensation costs, medical costs, insurance premiums, and lost and restricted workdays.

Whenever possible, devices should be used to help with patient transfers. Various devices such as draw sheets, slide boards, rollers, slings, belts, and mechanical or electronic hoists (to lift the patient) have been designed to assist healthcare workers and patients. The main lesson to be learned from studies about such devices is that each home situation must be assessed separately to find out which device will be the most suitable for (1) the persons using it, (2) the place(s) it will be used, and (3) the task(s) for which it will be used [Garg and Owen 1992; Zelenka et al. 1996; Elford et al. 2000]. Recognizing the importance of ergonomics for protecting the safety of healthcare workers, the Occupational Safety and Health Administration (OSHA) has issued ergonomics guidelines for nursing homes that emphasize the proper use of assistive devices during patient handling [OSHA 2003]. In addition, the VISN 8 Patient Safety Center of Inquiry [2007] has published a resource guide about safe patient handling and movement. The guide describes assistive devices and elements of an ergonomics program that have been tested within the Veterans' Health Administration and are being used on an ongoing basis at many other inpatient healthcare facilities. Some of the information from these sources is specific to nursing homes and hospitals, yet much of it applies to home healthcare. Parsons et al. [2006 a,b] has written two articles specifically about preventing musculoskeletal disorders in home healthcare workers.

Figures 2.1 through 2.10 provide examples of assistive devices that can be used in home settings. Many more types of products designed for a variety of patient-handling and other home healthcare needs are commercially available. Patients, family members, and home healthcare workers should consult with equipment vendors and the patient's primary doctor to select proper assistive devices that will lessen the worker's strain without decreasing the patient's safety or comfort. In some cases, a prescription is required to get such devices. Generally, a patient's insurance at least partially covers the costs. It's most important that all persons who use a lifting device be fully trained to use it safely. Periodic maintenance and cleaning for some devices, such as hoists, are required.

2.5 What can I do to prevent musculoskeletal disorders?

Some simple solutions have greatly reduced the number of patient transfers that nursing personnel need to perform. For example, Garg and Owen [1992] found that using a hoist with a built-in weighing scale eliminated transfers for the sole purpose of weighing the patient (from wheelchair to weighing scale and from weighing scale to wheelchair) and using a rolling toileting or showering chair reduced the six transfers needed for toileting and showering (bed to wheelchair, wheelchair to toilet, toilet to wheelchair, wheelchair to bathtub, bathtub to wheelchair, and wheelchair to bed) to two transfers (bed to toileting/showering chair and toileting/showering chair to bed).

Equipment such as adjustable beds, raised toilet seats, shower chairs, and grab bars are also helpful for reducing musculoskeletal risk factors. This type of equipment keeps the patient at an acceptable lift height and allows the patient to help himself or herself during transfer when possible.

Even when assistive devices are used during patient care, it is impossible to completely eliminate the need for some amount of physical exertion. For example, when using a hoist, the healthcare worker must move the patient in order to fasten the sling, and workers must support and balance the patient while using hoists and other devices. These tasks will always pose some risk of injury [VISN 8 Patient Safety Center of Inquiry 2007]. To lessen the risk, certain principles of body mechanics should be followed as much as possible to avoid harmful postures [Owen and Garg 1990; Zhuang et al. 1999; Garg and Owen 1992; Nelson et al.1997; Nelson et al. 2003]. Some strategies for effective body mechanics in patient handling are described in the Recommendations for Workers.

2.5.1 Recommendations for Employers

- Consult with a professional with expertise in patient-care ergonomics to determine when assistive devices are necessary and to provide training on proper use of the equipment.
- Provide ergonomic training for workers.
- Evaluate each patient-care plan to determine whether ergonomic assistive devices are appropriate.

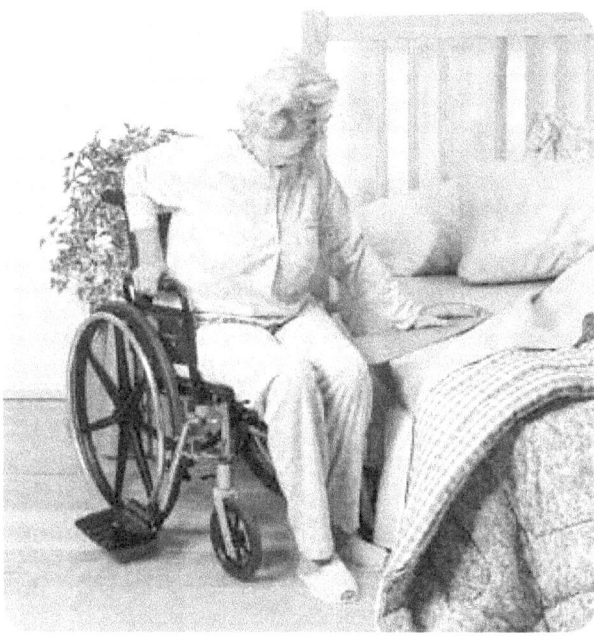

Figure 2.1. Slide/tranfer board (Copyright by Sammons Preston Rolyan. Reprinted with permission.)

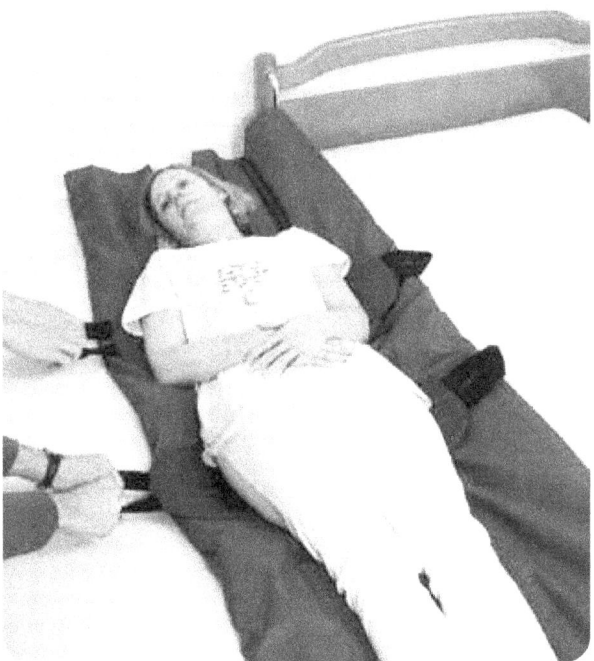

Figure 2.2. Slide/draw sheet (Copyright by SureHands Lift and Care Systems. Reprinted with permission.)

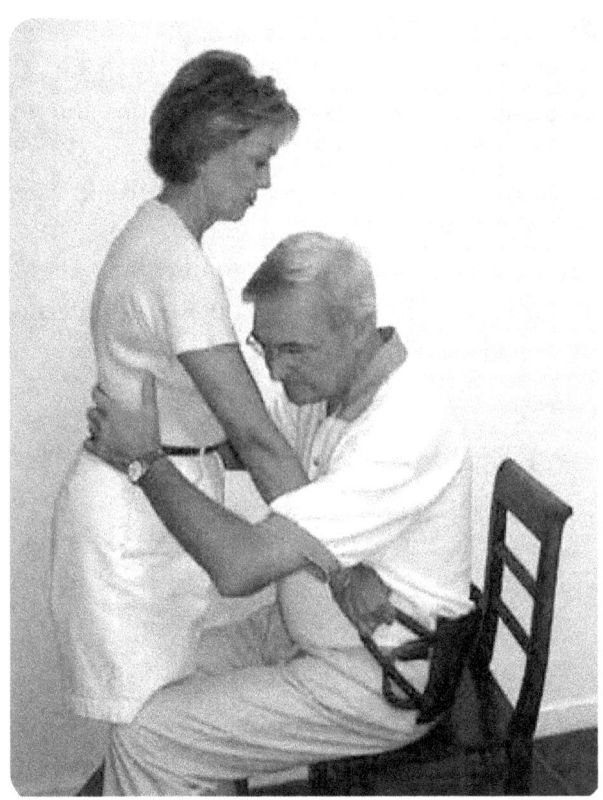

Figure 2.3. Patient moving sling (Copyright by Sammons Preston Rolyan. Reprinted with permission.)

Figure 2.4. Rolling toilet/shower chair (Copyright by Sammons Preston Rolyan. Reprinted with permission.)

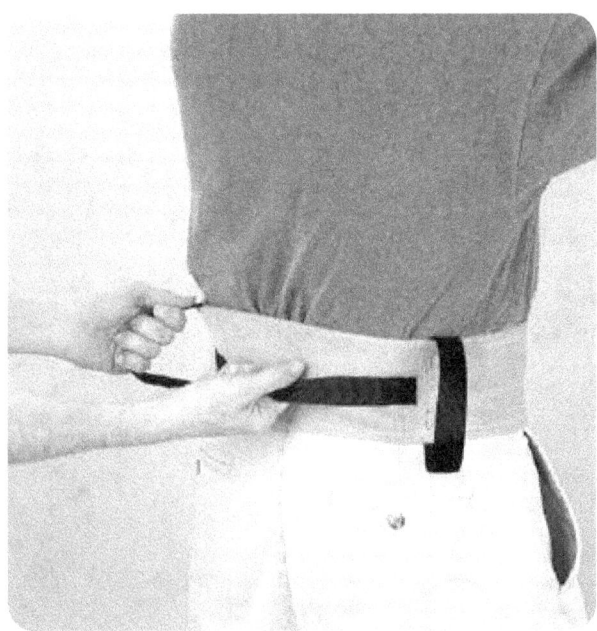

Figure 2.5. Gait/walking belt (Copyright by Sammons Preston Rolyan. Reprinted with permission.)

Figure 2.6. Stationary shower chair (Copyright by Sammons Preston Rolyan. Reprinted with permission.)

2 • Musculoskeletal Disorders and Ergonomic Interventions

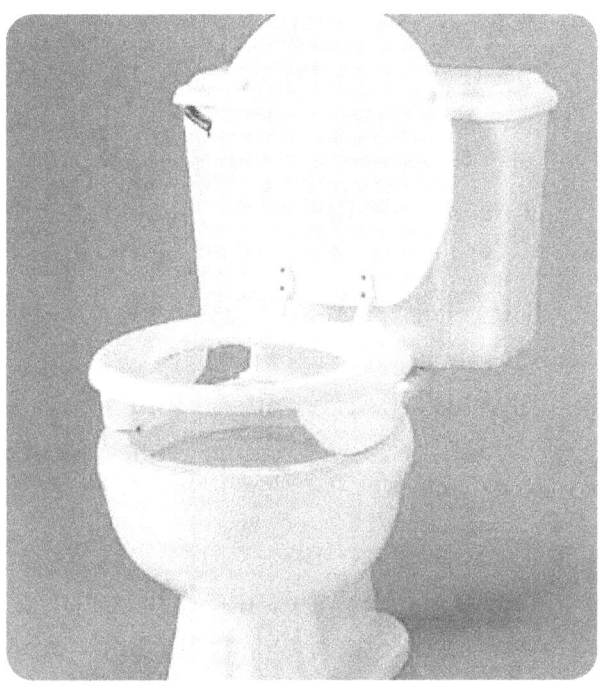

Figure 2.7. Raised toilet seat (Copyright by Sammons Preston Rolyan. Reprinted with permission.)

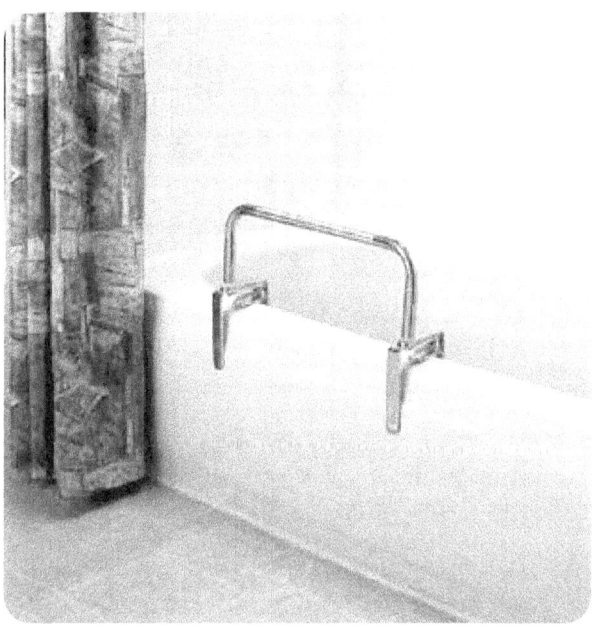

Figure 2.8. Grab bars (Copyright by Sammons Preston Rolyan. Reprinted with permission.)

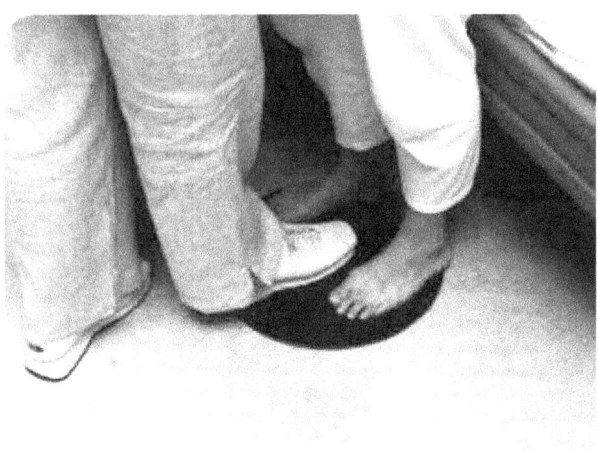

Figure 2.9. Rotation disk (Copyright by Sure Hands Lift and Care Systems. Reprinted with permission.)

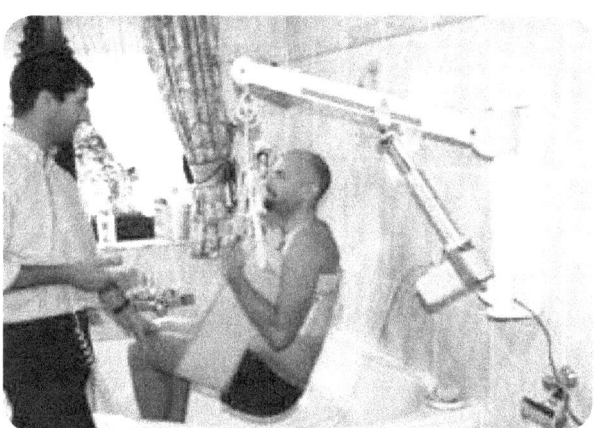

Figure 2.10. Wall sling (Copyright by Sure Hands Lift and Care Systems. Reprinted with permission.)

- Provide ergonomic assistive devices when needed.
- Reassess the training, the care plan, and the assistive devices once installed and in use by the caregiver.

Bringing ergonomic approaches into home healthcare settings is challenging because of the following:

- Workers may think assistive devices will be difficult to work with and time-consuming.
- Patients and family caregivers may fear that assistive devices will be unsafe or uncomfortable.
- Patients and families may be unwilling or unable to accept changes in the home.
- A device may be too expensive for the patient and family.

If patients and families are resistant to installing or buying an assistive device, the employers should inform them about the risks involved in moving patients when a device is not used. These risks may include the following:

- An overexerted worker could accidentally harm the patient.
- The patient may be injured by being dropped, jared, or not properly handled during unassisted transfers.

2.5.2 Recommendations for Workers

- Use ergonomic assistive devices if they are available.
- Move along the side of the patient's bed to stay in safe postures while performing tasks at the bedside. Do not stand in one location while bending, twisting, and reaching to perform tasks.
- When you are manually moving the patient, stand as close as possible to the patient without twisting your back, keeping your knees bent and feet apart. To avoid rotating the spine, make sure one foot is in the direction of the move.
- Use a friction-reducing device such as a slip sheet whenever possible [Nelson et al. 2003]. Using gentle rocking motions can also reduce exertion while moving a patient.
- Pulling a patient up in bed is easier when the head of the bed is flat or down. Raising the patient's knees and encouraging the patient to push (if possible) can also help.
- Apply anti-embolism stockings by pushing them on while standing at the foot of the bed. This position reduces exertion compared with standing at the side of the bed.

Notify your employer if you feel you would benefit from additional training or ergonomic assistive devices.

[Owen and Garg 1990; Zhuang et al. 1999; Garg and Owen 1992; Nelson et al.1997; Nelson et al. 2003]

2.6 Resources

CDC. Preventing falls among seniors (topic page) [www.cdc.gov/ncipc/duip/spotlite/fallpub.htm].

NIOSH [2006]. Safe lifting and movement of nursing home residents. U.S. Department of Health and Human Services, Centers for Disease Control and Prevention, National Institute for Occupational Safety and Health, DHHS (NIOSH) Publication No. 2006–117 [www.cdc.gov./niosh/docs/2006-117/].

OSHA. Healthcare wide hazards module—ergonomics [www.osha.gov/SLTC/etools/hospital/hazards/ergo/ergo.html].

2.7 References

BLS [2008a]. Table R5. Incidence rates for nonfatal occupational injuries and illnesses involving days away from work per 10,000 full-time workers by industry and selected natures of injury or illness, 2007. Washington, DC: U.S. Department of Labor, Bureau of Labor Statistics [www.bls.gov/iif/oshwc/osh/case/ostb1947.txt].

BLS [2008 b]. Table R8. Incidence rates for nonfatal occupational injuries and illnesses involving days away from work per 10,000 full-time workers by industry and selected events or exposures leading to injury or illness, 2007. Washington, DC: U.S. Department of Labor, Bureau of Labor Statistics [www.bls.gov/iif/oshwc/osh/case/ostb1950.txt].

Brulin C, Goine H, Edlund C, Knutsson A [1998a]. Prevalence of long-term sick leave among female home care personnel in northern Sweden. J Occup Rehab 8(2):103–111.

Brulin C, Gerdle B, Granlund B, Hoog J, Knutson A, Sundelin G [1998b]. Physical and psychosocial work-related risk factors associated with musculoskeletal symptoms among home care personnel. Scand J Carin Sci 12:104–110.

Chhokar R, Engst C, Miller A, Robinson D, Tate R, Yassi A [2005]. The three-year economic benefits of a ceiling lift intervention aimed to reduce healthcare worker injuries. Appl Ergon 36:223–229.

Collins J, Wolf L, Bell J, Evanoff B [2004]. An evaluation of a "best practices" musculoskeletal injury prevention program in nursing homes. Inj Prev 10(4):206–211.

Dellve L, Lagerstrom M, Hagberg M [2003]. Work-system risk factors for permanent work disability among home-care workers: a case-control study. Int Arch Occup Environ Health 76(3):216–224.

El-Askari E and DeBaun B [1999]. The occupational hazards of home health care. In Charney W., Fragula G. eds. The epidemic of health care worker injury: an epidemiology. Boca Ratonm FL: CRC Press, pp. 201–213.

Elert J, Brulin C, Gerdle B, Johansson H [1992]. Mechanical performance level of continuous contraction and muscle pain symptoms in home care personnel. Scand J Rehab Med 24:141–151.

Elford W, Straker L, Strauss G [2000]. Patient handling with and without slings: an analysis of the risk of injury to the lumbar spine. Appl Ergonomics 31:185–200.

Engst C, Chhokar R, Miller A, Tate R, Yassi A [2005]. Effectiveness of overhead lifting devices in reducing the risk of injury to care staff in extended care facilities. Ergonomics 48:187–199.

Evanoff B, Wolf L, Aton E, Canos J, Collins J [2003]. Reduction in injury rates in nursing personnel through introduction of mechanical lifts in the workplace. Am J Ind Med 44:451–457.

Fujishiro K, Weaver J, Heaney C, Hamrick C, Marras W [2005]. The effect of ergonomic interventions in healthcare facilities on musculoskeletal disorders. Am J Ind Med 48:338–347.

Galinsky T, Waters T, Malit B [2001]. Overexertion injuries in home health care workers and the need for ergonomics. Home Health Care Serv Q 20(3):57–73.

Garg A, Owen B [1992]. Reducing back stress to nursing personnel: an ergonomic intervention in a nursing home. Ergonomics 35:1353–1375.

Hignett S [2003]. Intervention strategies to reduce musculoskeletal injuries associated with handling patients: a systematic review. Occup Environ Med 60(9):E6.

Jarrell RB [1997]. Home care workers: injury prevention through risk factor reduction. Occup Med State of the Art Reviews 12(4):757–766.

Johansson J [1995]. Psychosocial work factors, physical work load and associated musculoskeletal symptoms among home care workers. Scand J Psychol 36:113–129.

Knibbe J, Friele R [1996]. Prevalence of back pain and characteristics of the physical workload of community nurses. Ergonomics 39(2):186–198.

Marras W, Davis K, Kirking B, Bertsche P [1999]. A comprehensive analysis of low-back disorder risk and spinal loading during the transferring and repositioning of patients using different techniques. Ergonomics 42(7):904–926.

Meyer J, Muntaner C [1999]. Injuries in home health care workers: an analysis of occupational morbidity from a state compensation database. Am J Ind Med 35:295–301.

Moens G, Dohogne T, Jacques P [1994]. Occupation and the prevalence of back pain among employees in health care. Arch Public Health 52:189–201.

Myers A, Jensen R, Nestor D, Rattiner J [1993]. Low back injuries among home health aides compared with hospital nursing aides. Home Health Care Serv Q 14(2/3):149–155.

Nelson A, Gross C, Lloyd J [1997]. Preventing musculoskeletal injuries in nurses: directions for future research. Sci Nursing 14(2):45–51.

Nelson A, Lloyd J, Menzel N, Gross C [2003]. Preventing nursing back injuries: Redesigning patient handling tasks. AAOHN J 51(3):126–134.

Nelson A, Matz M, Chen F, Siddharthan K, Lloyd J, Fragala G [2006]. Development and evaluation of a multifaceted ergonomics program to prevent injuries associated with patient handling tasks. Int J Nurs Stud 43:717–733.

Nelson A, Collins J, Siddharthen K, Matz M, Waters T [2008]. Link between safe patient handling and patient outcomes in long-term care. Rehabil Nurs 33:33–43.

NIOSH [1997]. Musculoskeletal disorders and workplace factors. A critical review of epidemiologic evidence for work-related musculoskeletal disorders of the neck, upper extremity, and low back. Cincinnati, OH: U.S. Department of Health and Human Services, Centers for Disease Control and Prevention, National Institute for Occupational Safety and Health, DHHS (NIOSH) Publication No. 97–141.

NIOSH [1999]. Long-term effectiveness of "zero-lift programs" in seven nursing homes and one hospital. By Garg A. Cincinnati, OH: U.S. Department of Health and Human Services, Centers for Disease Control and Prevention, National Institute for Occupational Safety and Health, NIOSH Contract Report No. U60/CCU512089–02.

NIOSH [2004]. Health hazard evaluation and technical assistance report: Alameda County Public Authority for In-Home Support Services, Alameda California. By Baron S, Habes D. Cincinnati, OH: U.S. Department of Health and Human Services, Centers for Disease Control and Prevention, National Institute for Occupational Safety Health, NIOSH HETA Report No. 2001–0139–2930.

Ono Y, Lagerstrom M, Hagberg M, Linden A, Malker B [1995]. Reports of work related musculoskeletal injury among home care service workers compared with nursery school workers and the general population of employed women in Sweden. Occup Environ Med 52:686–693.

Orr GB [1997]. Ergonomics programs for health care organizations. Occup Med 12(4):687–700.

OSHA [2003]. Ergonomics: guidelines for nursing homes Washington, D.C. U.S. Department of Labor: Occupational Safety and Health Administration [www.osha.gov/ergonomics/guidelines/nursinghome/index.html].

Owen B [1999]. The epidemic of back injuries in health care workers in the U.S. In: Charney W, Fragala G, eds. The epidemic of health care worker injury: an epidemiology. Boca Raton, FL: CRC Press LLC, pp. 47–56.

Owen B [2003]. Decreasing back stress in home care. Home Healthc Nurse 21(3):180–186.

Owen B, Garg A [1990] Assistive devices for use with patient handling tasks. In: Das B, ed. Advances in industrial ergonomics and safety. Philadelphia, PA: Taylor & Francis.

Owen B, Garg A [1991]. Reducing risk for back pain in nursing personnel. AAOHN J 39(1):24–33.

Parsons K, Galinsky T, Waters T [2006a]. Suggestions for preventing musculoskeletal disorders in home health care workers Part 1. Home Healthc Nurse 24(3):159–164.

Parsons K, Galinsky T, Waters T [2006b]. Suggestions for preventing musculoskeletal disorders in home health care workers Part 2. Home Healthc Nurse 24(4):227–233.

Pohjonen T, Punakallio A, Louhevaara V [1998]. Participatory ergonomics for reducing load and strain in home care work. Int J Ind Ergonomics 21:345–352.

Pohjonen T [2001]. Age-related physical fitness and the predictive values of fitness tests for work ability in home care work. J Occup Environ Med 43(8):723–730.

Ronald L, Yassi A, Spiegel J, Tate R, Tait D, Mozel M [2002]. Effectiveness of installing overhead ceiling lifts: Reducing musculoskeletal injuries in an extended care hospital unit. AAOHN J 50(3):120–127.

Santaguida P, Pierrynowski M, Goldsmith C, Fernie G [2005]. Comparison of cumulative low back loads of caregivers when transferring patients using overhead and floor mechanical lifting devices. Clinical Biomech 20:906–916.

Spiegel J, Yassi A, Ronald L, Tate R, Hacking P, Colby T [2002]. Implementing a resident lifting system in an extended care hospital: demonstrating cost-benefit. Am Assoc Occup Health Nurs 50:128–134.

Torgen M, Nygard C-H, Kilbom A [1995]. Physical work load, physical capacity and strain among elderly female aides in home-care service. Eur J Appl Physiol 71:444–452.

VISN 8 Patient Safety Center of Inquiry [2007]. Resource guide: safe patient handling and movement [www.visn8.med.va.gov/patientsafetycenter/safePtHandling/default.asp].

Waters T, Collins J, Galinsky T, Caruso C [2006]. NIOSH research efforts to prevent musculoskeletal disorders in the healthcare industry. Orthop Nurs 25:380–389.

Waters T, Putz-Anderson V, Garg A, Fine L [1993]. Revised NIOSH equation for the design and evaluation of manual lifting tasks. Ergonomics 36:749–776.

Zelenka J, Floren A, Jordan J [1996]. Minimal forces to move patients. Am J Occup Ther 50(5):354–361.

Zhuang Z, Stobbe T, Hsiao H, Collins J, Hobbs G [1999]. Biomechanical evaluation of assistive devices for transferring residents. Appl Ergonomics 30:285–294.

Latex Allergy

3.1 Introduction

A NIOSH Alert, *Preventing Allergic Reactions to Natural Rubber Latex in the Workplace* [NIOSH 1997], outlines many of the safety and health issues related to occupational exposure to products that contain natural rubber latex (NRL). This chapter includes information from the Alert as well as from other material useful to healthcare workers. Unless otherwise cited, the material in this chapter is from the Alert.

In this chapter, latex means NRL and includes products made from dry, natural rubber. Allergic reactions to latex range from mild to severe, including skin rashes; hives; nasal, eye, or sinus symptoms; asthma; and (rarely) shock. Most persons who are sensitive to latex are not born with the allergy. They develop it after repeated exposures to products that contain latex. Limiting exposure to latex is important for both home healthcare workers and the patients in their care to prevent allergic reactions to latex.

3.2 What are some sources of exposure to latex?

Although many different products (see Tables 1 and 2) may expose workers in different professions to latex, workers in the healthcare industry are frequently affected because of their repeated exposure: commonly wearing latex gloves [Liss and Sussman 1999] and using latex-containing medical equipment. Gloves made from latex are still used because of their low cost, tactile qualities, durability, and resistance to leakage [Stehlin 1992; Hunt et al. 1996; Douglas et al. 1997]. Some latex gloves contain a powder that is used as a lubricant, and the proteins responsible for latex allergy attach to this powder. When powdered gloves are worn, more protein reaches the skin, and when these gloves are changed, the particles of powder are released into the air and are inhaled. Therefore, the use of powder-free gloves may decrease both skin and respiratory exposure to latex [Allmers et al. 1998]. Also, using non powdered latex gloves with reduced protein lowers allergen exposure and has been shown to decrease the prevalence of latex reactions in hospital settings [Allmers et al. 1998; Turjanmaa et al. 2000].

3.3 What are the effects of latex exposure?

Three types of reactions can occur in persons using latex products:

- Irritant contact dermatitis
- Allergic contact dermatitis (delayed hypersensitivity)
- Latex allergy (immediate hypersensitivity)

3.3.1 What is irritant contact dermatitis?

Irritant contact dermatitis is the most common adverse reaction associated with using

Table 1. Medical and dental products that may contain latex

Adhesive tape	Anesthesia masks	Bite blocks	Blood pressure cuffs
Catheters	Certain epidural catheter injection adapters	Condom urinary collection devices	Dental dams
Elastic bandages	Electrode pads	Endotracheal tubes	Enema tubing tips
Goggles	Hemodialysis equipment	Injection ports	Intravenous tubing
Latex cuffs on plastic tracheal tubes	Oral and nasal airways	Reservoir breathing bags	Respiratory protective masks
Rubber aprons	Rubber tops of multidose vials	Rubber ventilator hoses/bellows	Stethoscopes
Stomach and intestinal tubes	Surgical and examination gloves	Surgical masks	Syringes
Teeth protectors	Tourniquets	Urinary catheters	Wound drains

Table 2. Household and office objects that may contain latex

Automotive tires	Baby bottle nipples	Balloons	Carpeting
Condoms	Diaphragms	Dishwashing gloves	Erasers
Expandable fabrics	Hot water bottles	Motorcycle and bicycle handgrips	Pacifiers
Racquet handles	Rubber bands	Shoe soles	Swimming goggles

latex gloves. Dry, itchy, irritated areas of the skin—most frequently the hands—are the symptoms [Sussman and Beezhold 1995]. Irritant contact dermatitis is not an allergy but a reaction to repeated exposure to an irritating substance. This skin condition can be caused by putting on and taking off latex gloves or gloves of other materials. It can also be caused by repeated hand washing and drying, incomplete hand drying, using cleaners and sanitizers, and repeated contact with powders added to some latex gloves. A skin rash may also be a first sign of latex allergy and of more significant reactions that may occur with continued exposure to latex.

3.3.2 What is allergic contact dermatitis?

Allergic contact dermatitis is caused by contact with chemicals added during harvesting, processing, or manufacturing latex products. This is a skin reaction that resembles the rash that occurs after contact with poison ivy. This rash, when caused by latex gloves, generally begins 24–96 hours after contact and may develop to oozing blisters or spread from the initial area of contact [Sussman and Beezhold 1995; NIOSH 1997].

3.3.3 What is latex allergy?

Latex allergy is potentially a more serious reaction than irritant contact or allergic contact dermatitis. The reaction may occur at low exposures if the person is highly sensitized. Although reactions usually occur within minutes of exposure, the symptoms may be delayed for a few hours. Mild reactions consist of redness of the skin, hives, or itching. More serious reactions might include runny nose, sneezing, itchy eyes, scratchy throat, and asthma (difficulty breathing, wheezing, and cough). Rarely, shock may occur, but a life-threatening reaction is seldom the first sign of latex allergy [NIOSH 1997].

A latex-exposed worker who develops any of the more serious allergic reactions given above, including unexplained shock, should be taken to a doctor right away. The doctor should ask the worker's medical history and may give a physical exam and medical testing. FDA-approved skin and blood tests are available. Occasionally, tests do not confirm a suspected latex allergy in someone who has a true latex allergy or may indicate allergy in someone without a compatible medical history. Therefore, clinical judgment from the doctor is important.

3.4 What are some products that contain latex ?

The preceding two tables list products that may contain latex. The tables are not complete lists; other products may contain latex [Stehlin 1992; NIOSH 1997]. The American Latex Allergy Association maintains lists of latex-free medical, dental, and consumer products that may be considered for substitution.

The FDA requires all natural rubber products that come in contact with humans be labeled to say that the products contain natural rubber latex and may cause allergic reactions [62 Fed. Reg.* 51021 (1997)], therefore any glove that contains latex will state so on the box.

3.5 How can I prevent exposure to latex?

The following recommendations can reduce or prevent exposure to latex [Sussman et al. 1994; Hunt et al. 1996; NIOSH 1997].

3.5.1 Recommendations for Employers

- Provide workers with nonlatex gloves when there is little contact with infectious materials.

- If the potential exists for contact with infectious materials, select gloves that pass the ASTM F1671 penetration test for resistance to bloodborne pathogens [Sustainable Hospitals 2007]. Various manufacturers of vinyl, nitrile, polymer, and latex gloves have appropriate gloves for infectious materials.

- If latex gloves are selected, provide reduced-protein, powder-free gloves.

- Provide training to supervisors and staff on latex allergy.

- Promptly arrange a medical evaluation for workers with early symptoms.

- Evaluate current prevention strategies whenever a worker is diagnosed with latex allergy.

- Frequently clean areas possibly contaminated with latex dust (upholstery, carpets, ventilation ducts, and plenums) in a manner that minimizes dust dispersal, such as use of a vacuum with a high-efficiency particulate air filter.

---

*Federal Register. See Fed. Reg. in references.

3.5.2 Recommendations for Workers

- Use nonlatex gloves for activities that are likely not to involve contact with infectious materials.
- Ask your employer for gloves that do not contain latex but still offer protection against infectious materials.
- If your employer supplies latex gloves, ask for reduced-protein, powder-free ones. These gloves may reduce the risk of latex allergy.
- Avoid oil-based creams or lotions when using latex gloves. Oil-based creams or lotions may cause the gloves to break down and deteriorate.
- Wash hands with a mild soap and dry hands completely after using gloves.
- Participate in training provided by your employer. Learn ways to prevent latex allergy.
- Recognize symptoms of latex allergy (rash; hives; flushing; itching; nasal, eye, and sinus irritation; asthma; and shock).
- If you develop symptoms of latex allergy, avoid direct contact with latex gloves and other latex-containing products until you can see a doctor. Until your appointment, also avoid areas where you may contact powder from latex gloves.
- If you are diagnosed with latex allergy, do the following:
    - Avoid touching, using, or being near latex-containing products.
    - Avoid areas where latex is likely to be inhaled (for example, where powdered latex gloves are being used).
    - Inform your employer and your personal healthcare professionals that you have latex allergy.
    - Wear a medical alert bracelet.
    - Follow your doctor's recommendations about latex allergy.
    - Before receiving any shots (such as the flu shot), be sure the person giving it uses a latex-free vial stopper [Primeau et al. 2001].
    - Before receiving a medical procedure or surgery, consult the specialist who will perform the procedure about any modifications that may be needed in the materials that will be used.

3.6 Resources

American Latex Allergy Association
3791 Sherman Road
Slinger, WI 53086
1-888-972-5378
[www.latexallergyresources.org/].

Canadian Society of Allergy and Clinical Immunology. Natural rubber latex allergy: a guideline for allergic patients [http://www.allergyfoundation.ca/website/latex_allergy_guidelines.htm].

NIOSH. Latex allergy: a prevention guide [www.cdc.gov/niosh/98-113.html].

NIOSH. Occupational latex allergies topic page [http://www.cdc.gov/niosh/topics/latex/].

Sustainable Hospitals. Alternative products and procedures [www.sustainablehospitals.org/HTMLSrc/Alternative.html].

3.7 References

Allmers H, Brehler R, Chen Z, Raulf-Heimsoth M, Fels H, Baur X [1998]. Reduction of latex aeroallergens and latex-specific IgE antibodies in sensitized workers after removal of powdered natural rubber latex gloves in a hospital. J Allergy Clin Immunol 101:171–178.

Douglas A, Simon TR, Goddard M [1997]. Barrier durability of latex and vinyl medical gloves in clinical settings. Am Ind Hyg Assoc J 58:672–676.

62 Fed. Reg. 51021 [1997]. Food and Drug Administration: Natural rubber-containing medical devices; user labeling. (Codified at 21 CFR 801.)

Hunt LW, Boone-Orke JL, Fransway AF, Fremstad CE, Jones RT, Swanson MC, McEvoy MT, Miller LK, Majerus ET, Luker PA, Scheppmann DL, Webb MJ, Yunginger JW [1996]. A medical-center-wide, multidisciplinary approach to the problem of natural rubber latex allergy. J Occup Environ Med 38(8):765–770.

Liss GM, Sussman GL [1999]. Latex sensitization: occupational versus general population prevalence rates. Am J Ind Med 35:196–200.

NIOSH [1997]. NIOSH alert: preventing allergic reactions to natural rubber latex in the workplace. Cincinnati, OH: U.S. Department of Health and Human Services, Centers for Disease Control and Prevention, National Institute for Occupational Safety and Health, DHHS (NIOSH) Publication No. 97–135.

Primeau M-N, Adkinson NF, Hamilton RG [2001]. Natural rubber pharmaceutical vial enclosures release latex allergens that produce skin reactions. J Allergy and Clin Immunol 107:958–962.

Stehlin D [1992]. When rubber rubs the wrong way. FDA Consum September; 26(7):16–21.

Sussman G, Beezhold DH [1995]. Allergy to latex rubber. Ann Intern Med 122:43–46.

Sussman G, Drouin MA, Hargreave FE, Douglas A, Turjanmaa K [1994]. Natural rubber latex allergy: a guideline for allergic patients. Canadian Society of Allergy and Clinical Immunology (CSACI).

Sustainable Hospitals [2007]. Alternative products and procedures. Lowell, MA: University of Massachusetts, Department of Work Environment [www.sustainablehospitals.org/HTMLSrc/Alternative.html].

Turjanmaa K, Reinikka-Railo H, Reunala T, Palosuo T [2000]. Continued decrease in natural rubber latex (NRL) allergen levels of medical gloves in nationwide market surveys in Finland and co-occurring decrease in NRL allergy prevalence in a large university hospital. J Clin Allergy Clin Immunol 104:S373.

Exposure to Bloodborne Pathogens and Needlestick Injuries

4.1 Introduction

Needlestick and other sharps injuries are a serious hazard in any medical care situation. These injuries are caused by different types of needles and sharps, such as scalpels and broken glass containers. Contaminated needles and sharps may inject healthcare workers with blood that contains pathogens such as hepatitis B virus (HBV), hepatitis C virus (HCV), and human immunodeficiency virus (HIV), all of which pose a grave, potentially lethal, risk. Although immunization is available to prevent hepatitis B illness, no immunization is available to prevent HCV or HIV. Preventing injuries from sharps and needlesticks is key to reducing potential exposures to bloodborne pathogens in home healthcare settings.

4.2 How serious is the risk of exposure from needlestick and sharps injuries?

It is estimated that 385,000 to 800,000 needlestick and other sharps injuries occur annually in all settings, but about half of these are not reported [Henry and Campbell 1995; CDC 1997; EPINet 1999; Osborn et al. 1999; CDC 2004]. Home healthcare workers give various reasons for not reporting such injuries: time-consuming post-injury process; anxiety surrounding the post-injury process; fear of being blamed as careless or thought of as a bad nurse by the employer; disease history of a patient (that is, patient thought not to be an infection risk); or fear of implications for present or future job prospects [Markkanen et al. 2007].

Activities associated with needlestick injuries include the following:

- Handling needles that must be taken apart or manipulated after use
- Disposing of needles attached to tubing
- Manipulating the needle in the patient
- Recapping needle
- Transferring body fluid between containers using needles or glass equipment
- Failing to dispose of used needles in puncture-resistant sharps containers
- Lack of proper workstations for procedures using sharps
- Rapid work pace and productivity pressures
- Bumping into a needle, sharps, or a worker
- Inadequate staffing and poor leadership

[McCormick et al. 1991; Yassi and McGill 1991; Clarke et al. 2002; CDC 2004; Wilburn 2004].

Home healthcare workers are responsible for the use and disposal of sharps equipment that they use in the patient's home. However, the patient or family may not appropriately dispose of sharps, thus putting the worker at risk. The worker may find contaminated sharps on any surface in the home or in wastebaskets. Focus groups of home healthcare workers have reported that syringes and lancets are left uncovered in various places in the home [Markkanen et al. 2007]. The home healthcare worker, without access to a standard sharps disposal container, often uses whatever is available for disposal (for example, coffee cans, milk jugs) [Backinger and Koustenis 1994; Haiduven 2000].

Pets and children in the home may be a dangerous distraction, increasing the risk of needlestick injury [Charney and Fragala 1999; Haiduven 2000; Markkanen et al. 2007]. The patient or family members may also be disruptive.

Home healthcare workers may also be exposed to bloodborne pathogens from episodes of sudden profuse bleeding (for example, bleeding tumors and amputations) and tasks involving wound care [Markkhanen et al. 2007].

4.3 What regulations should I be aware of?

Federal legislation has shown an interest in preventing needlestick injuries and the diseases associated with needlestick injuries. The OSHA bloodborne pathogens standard [29 CFR* 1910.1030] is the Federal standard that protects workers against occupational exposures to bloodborne diseases. Since 1991 when the standard was first published, manufacturers have supplied new, safer designs for medical devices to reduce or eliminate needlesticks and other exposure incidents. OSHA updated the standard in 2001 with additional information about needleless systems, needle-containing equipment with safety features, and needlestick safety issues related to the OSHA bloodborne pathogens standard [56 Fed. Reg.† 2 64004 (2001)]. Employers and home healthcare workers are encouraged to visit the OSHA Web site (www.osha.gov) to obtain complete information about the bloodborne pathogens standard. Some of the requirements of the standard include the following:

- The employer must create a written exposure-control plan designed to eliminate or minimize worker exposure to bloodborne pathogens, and review it annually. The plan must include a determination of potential employee exposures for the workplace and a consideration of safe medical devices that may be newly available.

- Compliance with standard precautions (formerly known as universal precautions): an infection-control principle that treats all blood and other potentially infectious materials as infectious.

- Engineering controls and work practices to eliminate or minimize worker exposure and training in these controls and work practices. Engineering controls isolate or remove the bloodborne

---
*Code of Federal Regulations. See CFR in references.
†Federal Register. See Fed. Reg. in references.

pathogens hazard from the workplace and include

— Sharps disposal containers

— Self-sheathing needles

— Safer medical devices, such as sharps with engineered injury protection and needleless systems

- Input from nonmanagerial employees responsible for patient care in selecting engineering controls (for example, medical devices with safety features) and work practices. This must be documented in the written exposure-control plan.

- Prohibition of bending, recapping, or removing contaminated needles from the syringe unless there is no feasible alternative

- Proper disposal including use of the sharps disposal containers, not overfilling the containers, prohibition of shearing or breaking contaminated needles, and disposal that meets State and Federal medical waste requirements

- Personal protective equipment provided to employees at no cost to them

- Free hepatitis B vaccinations offered to workers with occupational exposure to bloodborne pathogens

- Post-exposure evaluation, with follow-up when appropriate

- Communication of hazards and training of workers

- Recordkeeping, including a sharps injury log maintained by the employer

- Protection of confidentiality of the injured worker in the injury log

- Procedures for evaluating circumstances surrounding exposure incidents

4.4 What about needleless systems and needle devices with safety features?

Evidence shows that using needleless systems or needle devices with safety features reduces needlestick injuries in I.V. systems and in relation to blood drawing [Gartner 1992; Yassi et al. 1995; Jagger 1996; CDC 1997; Lawrence et al. 1997; NCCC and DVA 1997; Zafar 1997; NIOSH 1998; CDC 2004].

4.4.1 What needleless systems and needle devices with safety features are available?

Below are examples of needleless systems and sharps with engineered injury protection:

- Needleless connectors for I.V.-delivery systems

- Protected needle I.V. connectors

- Needles that retract into a syringe or vacuum-tube holder (see Figure 4.1)

- Hinged or sliding shields attached to phlebotomy needles, winged-steel needles, and blood gas needles

- Protective encasements to receive an I.V. stylet as it is withdrawn from the catheter

- Sliding needle shields attached to disposable syringes and vacuum tube holders

- Self-blunting phlebotomy and winged-steel needles (see Figure 4.1)

- Retractable finger or heel-stick lancets (see Figure 4.2)

4 • *Exposure to Bloodborne Pathogens and Needlestick Injuries*

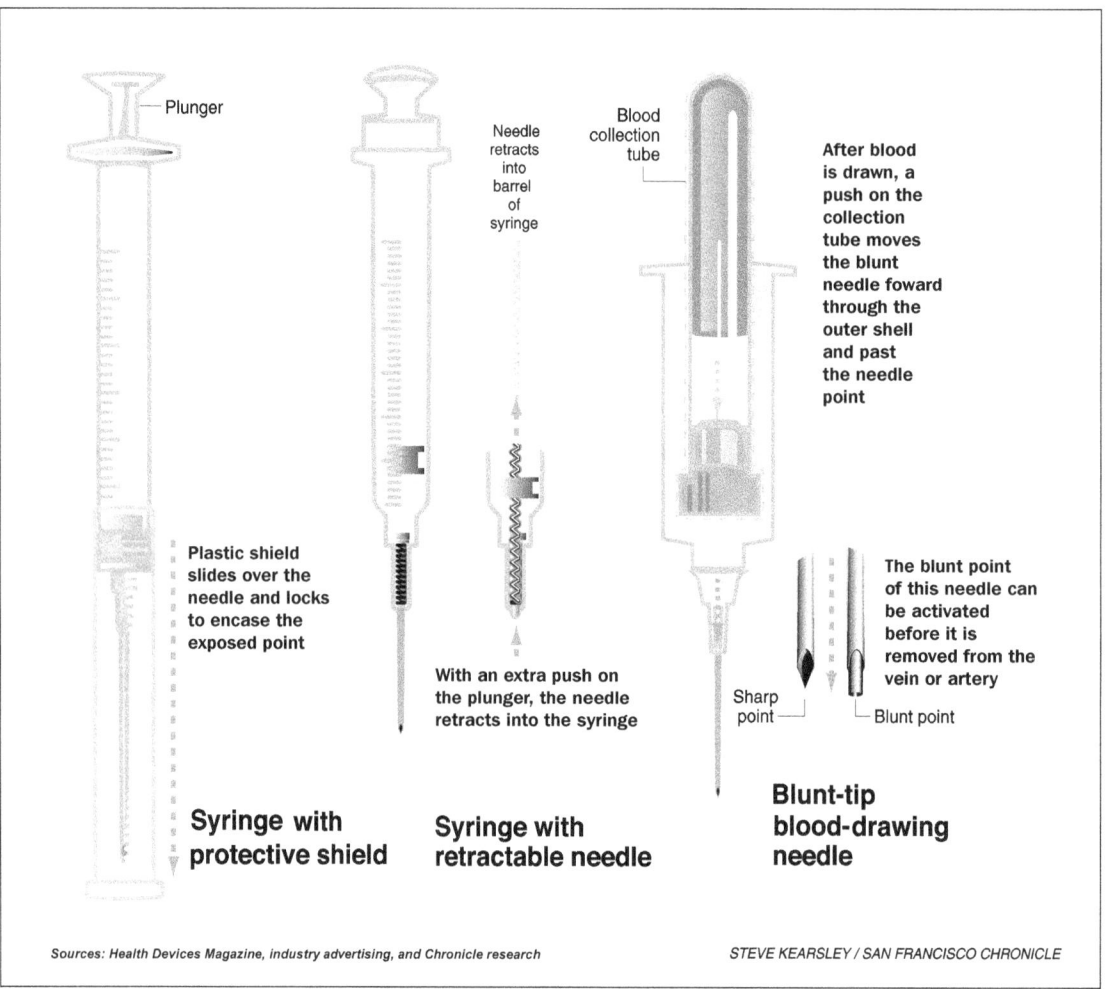

Figure 4.1. Three examples of syringes with safety features. (These drawings are presented for educational purposes and do not imply endorsement of a particular product by the National Institute for Occupational Safety and Health [NIOSH].)

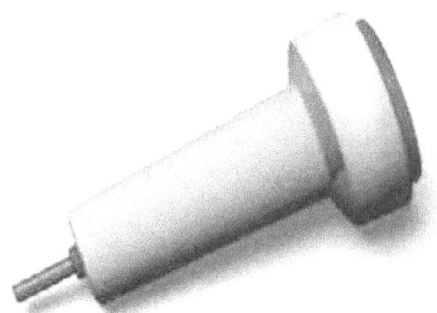

Figure 4.2. Example lancet with safety features. (This drawing is presented for educational purposes and does not imply endorsement of a particular product by the National Institute for Occupational Safety and Health [NIOSH].)

4.4.2 How do I select and evaluate needleless systems and needle devices with safety features?

Selecting and evaluating needle devices with safety features should include the following steps:

- Forming a multidisciplinary team to develop a plan to reduce needlestick injuries and evaluate needle devices with safety features

- Seeking input from, or including, non-managerial employees responsible for direct patient care and any other workers at risk of sharps injuries (The team should also participate in the implementation and evaluation of the plan that is developed.)

- Identifying whether and how needlestick injuries are occurring and how devices with safety features are being used

- Identifying needles or needleless devices with safety features that differ in design and features

- Performing visual and practical investigation of any design(s) selected

- Evaluating information (preferably from multiple sources) about the devices

- Evaluating the product(s) chosen, including input from workers who represent the range of potential users. The steps of the evaluation should include
  — establishing criteria to evaluate the device,
  — carrying out follow-up to obtain feedback, identify problems, and provide continued guidance, and
  — monitoring the use of a new device to determine any problems or whether further training is needed.

[NIOSH 1999; OSHA 2001; CDC 2004]

4.5 What can I do to prevent and control needlestick and sharps injuries?

4.5.1 Recommendations for Employers

- Provide a bloodborne pathogens program that meets all the requirements of the OSHA bloodborne pathogens standard (29 CFR 1910.1030).

- Eliminate the use of needle devices whenever safe and effective alternatives are available (for example, connecting parts of an I.V. system).

- Provide needle devices with safety features and determine which safety features are most effective and acceptable for tasks in the workplace (4.4.1).

- Establish an exposure-control plan; evaluate and update it annually.

- Analyze sharps-related injuries in the workplace to determine hazards and injury patterns. If patterns of injury develop, consider the following options:
  — Change work practices to decrease the specific activities associated with the injuries.
  — Train employees in new ways to do tasks that are known to have caused injury.
  — Use different needle devices than those associated with the injuries.

- Promote work practices that decrease the chance of a needlestick injury (for example, methods of transferring body fluids without the use of needles).

- Train workers in the safe use and disposal of all types of sharps and needle devices.

- Train workers to plan for unexpected movement and to watch for improperly disposed needles.

- Establish procedures and systems for the reporting, timely follow-up, and medical evaluation of all needlestick or sharps-related injuries.

- Establish a system to evaluate prevention efforts and provide feedback to workers and management.

- Provide standard-labeled, leak-proof, puncture-resistant sharps containers for workers to carry in their vehicles for use as needed when an adequate sharps container is not easily available in the home.

- Ensure that the patient or any other caregivers for the patient (for example, family members) receive training in infection control to help them understand and comply with the practices and precautions of the home healthcare worker [Valenti 1995].

- Provide post-exposure evaluation and follow-up, including post-exposure prophylaxis when appropriate.

4.5.2 Recommendations for Workers

- Participate in your employer's bloodborne pathogens program.

- Avoid using needles whenever safe and effective alternatives are available.

- Help your employer select and evaluate devices with safety features (see 4.4.1).

- Use devices with safety features provided by your employer.

- Refrain from recapping or bending contaminated needles.

- Before starting a procedure, plan for the safe handling and disposal of needles. Dispose of used needle devices and any potentially contaminated sharps materials promptly in designated sharps disposal containers.

- Carry standard-labeled, leak-proof, puncture-resistant, sharps containers with you to homes; do not assume the containers will be available in the home.

- Secure used sharps containers during transport to prevent spilling.

- Report any needlestick and other sharps injuries promptly to receive follow-up care.

- Follow standard precautions, infection prevention, and general hygiene practices consistently.

4.6 What should I do if I am exposed to the blood of a patient?

If you experienced a needlestick or sharps injury or were exposed to the blood or other body fluid of a patient during the course of your work, immediately follow these steps:

- Wash needlesticks and cuts with soap and water.

- Flush splashes to the nose, mouth, or skin with water.

- Irrigate eyes with clean water, saline, or sterile irrigants.

- Report the incident to your supervisor.
- Immediately seek medical treatment.

4.7 Resources

CDC. Workbook for designing, implementing, and evaluating a sharps injury prevention program [www.cdc.gov/sharpsSafety/].

CDC. Viral hepatitis [www.cdc.gov/ncidod/diseases/hepatitis/index.htm].

CDC. Hospital infections [www.cdc.gov/ncidod/dhqp/].

ECRI. [https://www.ecri.org/Documents/Sharps_Safety/SSNP_toc.pdf].

NIOSH. Needlestick injuries and bloodborne infections diseases topic page [www.cdc.gov/niosh/topics/bbp/].

OSHA. OSHA Pub No. 3186, Model bloodborne pathogens exposure plan [www.osha.gov/Publications/osha3186.html].

The University of California, San Francisco, toll-free phone number for clinicians to call for advice on post-exposure prophylaxis: 1–888–448–4911.

California Department of Health Services
Occupational Health Branch
1515 Clay Street, Suite 1901
Oakland, CA 94612
[www.ucsf.edu/hivcntr/].

The University of Virginia International Health Care Workers Safety Center and the EPINet needlestick injury data collection system [www.healthsystem.virginia.edu/internet/epinet/about_epinet.cfm].

International Healthcare Worker Safety Center
Health Sciences Center, University of Virginia
Box 407
Charlottesville, VA 22908
[www.healthsystem.virginia.edu/internet/epinet/].

4.8 References

Backinger CL, Koustenis GH [1994]. Analysis of needlestick injuries to health care workers providing home care. Am J Infect Control 22:300–306.

CDC [1997]. Evaluation of safety devices for preventing percutaneous injuries among health-care workers during phlebotomy procedures—Minneapolis-St. Paul, New York City, and San Francisco, 1993–1995. MMWR 46(2):21–25.

CDC [2004] Workbook for designing, implementing, and evaluating a sharps injury prevention program. Atlanta, GA: U.S. Department of Health and Human Services, Centers for Disease Control and Prevention [http://www.cdc.gov/sharpssafety/].

CFR. Code of Federal Regulations. Washington, DC: U.S. Government Printing Office, Office of the Federal Register.

Charney W, Fragala G [1999]. The epidemic of health care worker injury: an epidemiology. Boca Raton, FL: CRC Press LLC, pp. 201–213.

Clarke SP, Sloane DM, Aiken LH [2002]. Effects of hospital staffing and organizational climate on needlestick injuries to nurses. Am J Pub Health 92(7):1115–1119.

EPINet [1999]. Exposure prevention information network data reports. Charlottesville, VA: University of Virginia, International Health Care Worker Safety Center.

56 Fed. Reg. 64004 [1991]. Occupational Safety and Health Administration: final rule on occupational exposure to bloodborne pathogens.

Gartner K [1992]. Impact of a needleless intravenous system in a university hospital. Am J Infect Control 20:75–79.

Haiduven D [2000]. Circumstances surrounding blood exposures and needle safety practices in home health care nurses [Dissertation]. San Francisco, CA: University of California.

Henry K, Campbell S [1995]. Needlestick/sharps injuries and HIV exposures among health care workers: national estimates based on a survey of U.S. hospitals. Minn Med 78:1765–1768.

Jagger J [1996]. Reducing occupational exposure to bloodborne pathogens: where do we stand a decade later? Infect Control Hosp Epidemiol 17(9):573–575.

Lawrence LW, Delclos GL, Felknor SA, Johnson PC, Frankowski RF, Cooper SP, Davidson A [1997]. The effectiveness of a needleless intravenous connection system: an assessment by injury rate and user satisfaction. Infect Control Hosp Epidemiol 18(3):175–182.

Markkanen P, Quinn M, Galligan C, Chalupka S, Davis L, Laramie A [2007]. There's no place like home: a qualitative study of the working conditions of home health care providers. JOEM 49:(3)327–337.

McCormick RD, Meisch MG, Ircink FG, Maki DG [1991]. Epidemiology of hospital sharps injuries: a 14-year prospective study in the pre-AIDS and AIDS eras. Am J Med 91(Suppl 3B):301S–307S.

NCCC, DVA [1997]. Needle stick prevention in the Department of Veterans Affairs; 1996 follow-up survey results. Milwaukee, WI: National Center for Cost Containment, Department of Veterans Affairs.

NIOSH [1998]. Selecting, evaluating, and using sharps disposal containers. Cincinnati, OH: U.S. Department of Health and Human Services, Centers for Disease Control and Prevention, National Institute for Occupational Safety and Health, DHHS (NIOSH) Publication No. 97–111.

NIOSH [1999]. Alert: preventing needlestick injuries in health care settings. Cincinnati, OH: U.S. Department of Health and Human Services, Centers for Disease Control and Prevention, National Institute for Occupational Safety and Health, DHHS (NIOSH) Publication No. 2000–108.

Osborne EHS, Papadakis MA, Gerberding JL [1999]. Occupational exposures to body fluids among medical students. A seven-year longitudinal study. Ann Intern Med 130(1):45–51.

OSHA [2001]. Bloodborne pathogens and needlestick prevention: OSHA standards. Washington, DC: U.S. Department of Labor, Occupational Safety and Health Administration [http://www.osha.gov/SLTC/bloodbornepathogens/standards.html l

Valenti WM [1995]. AIDS: Problem solving in infection control. Infection control, human immunodeficiency virus, and home health care: II. Risk to the caregiver. Am J Infect Control 23:8–81.

Wilburn S [2004]. Needlestick and sharps injury prevention. Online J Issues Nurs 9(3):5.

Yassi A, McGill ML [1991]. Determinants of blood and body fluid exposure in a large teaching hospital: hazards of the intermittent intravenous procedure. Am J Infect Control 19(3):129–135.

Yassi A, McGill ML, Khokhar JB [1995]. Efficacy and cost-effectiveness of a needleless intravenous access system. Am J Infect Control 239(2):57–64.

Zafar AB, Butler RC, Podgorny JM, Mennonna PA, Gaydos LA, Sandiford JA [1997]. Effect of a comprehensive program to reduce needlestick injuries. Infect Control Hosp Epidemiol 18(10):712–71.

Occupational Stress

5.1 Introduction

Home healthcare work involves challenges that are not present in hospital or other in-patient healthcare settings. Not many studies have looked into stress levels of home healthcare workers, but the few studies that have show that home healthcare may be quite stressful. The home setting may involve stressors, such as lack of control over work planning, that are risks for shoulder and neck pain, especially when combined with physical risk factors such as strenuous postures [Johansson 1995; Brulin et al. 1998a]. Attracting workers and retaining them is a high priority for many home healthcare agencies, and providing a more healthful, less stressful, work climate is an important part of any retention strategy. The following sections discuss job stressors present in home healthcare work and provide suggestions for how job stress may be prevented or reduced for home healthcare workers.

5.2 What are some specific stressors of home healthcare workers?

NIOSH defines job stress as "the harmful physical and emotional responses that occur when the requirements of the job do not match the capabilities, resources, or needs of the worker" [NIOSH 1999]. Job stressors include job and task demands such as work overload, time pressure, lack of task control and role ambiguity; and organizational factors, such as poor interpersonal relations, lack of support from supervisors and co-workers, and unfair management practices [Hurrell and Murphy 1992]. Other sources of stress, which may be of particular importance in the home healthcare environment, are socioeconomic factors, training and career development issues, and conflict between work and family roles and responsibilities [Sauter and Swanson 1996].

Home healthcare workers report some of the same stressors as other healthcare workers:

- Ill and dying clients [Davidhizar 1999]
- Workload and time pressures [Jarrell 1997]
- Increasing emphasis on healthcare cost savings [Davidhizar 1999]
- Patient aggression [El-Askari and DeBaun 1999]
- Patients who are disoriented, irritable, or uncooperative [BLS 2008]

In addition, home healthcare workers may have to deal with stressors that healthcare workers in hospitals or other inpatient healthcare settings do not: their work is not directly supervised, they generally work alone, they might travel through unsafe neighborhoods, and they may have to face alcohol or drug abusers, family arguments, dangerous dogs, or heavy traffic.

Employers may not take a proactive enough stance in removing workers from an unsafe work environment or in providing support when workers encounter abusive behavior from the client or the client's family [Kendra et al. 1996]. Families may expect more from home healthcare workers than their duties require them to provide. Workers may be unsure whose instructions they should follow: the client's or those of the agency that employs them [Prager 1996].

Home healthcare workers face time pressures arising from their client loads. Time pressure may reduce the level of service [Prager 1996]. Home healthcare workers report that they shorten their visits if they feel unsafe [Kendra et al. 1996]. Workers may have to deal with clients who do not comply with prescribed medicine orders or who refuse services [Kendra et al. 1996]. Home healthcare workers have reported an increase in paperwork per each client visit because of state and federal regulatory requirements [Davidhizar 1999].

Some studies suggest that home healthcare workers may have more on-the-job stress than other comparable jobs, like teachers and child care workers. Johansson [1995] found that, compared with teachers and child care workers, home healthcare workers reported having less control over and being less excited by their work. Home healthcare workers took the most long-term sick leave (30 days or more per year) and had the second highest frequency of absenteeism [Brulin et al. 1998b].

5.3 What can I do to prevent and control occupational stress?

Both employers and employees can take actions to reduce stress.

5.3.1 Recommendations for Employers

- Provide frequent, quality supervision and agency staff support.
- Provide adequate job training and preparation, including continuing education opportunities.
- Hold regular staff meetings in which problems, frustrations, and solutions can be discussed.
- Include lunch breaks and sufficient travel time in workers' schedules and allow self-paced work.
- Have policies and procedures in place to ensure worker safety [Kendra et al. 1996].
- Provide access to an employee assistance program or other means of counseling support.
- Provide wages and benefits that are competitive with what other service organizations are offering [Prager 1996; Jarrell 1997; Stonerock 1997].

That last recommendation is particularly important for retaining home healthcare workers. In a survey sample, Kennedy-Malone [1996] found that 50% of home healthcare workers stated that "no pay increase" was a "very important" reason that they may resign; 40% said the same for "no health insurance."

5.3.2 Recommendations for Workers

- Develop effective coping strategies; try to put a positive spin on things. For example, think of ways a stressful situation will help you become a better healthcare worker.

- Improve time management or planning skills through training your employer may provide [Davidhizar 1999].
- Perform relaxation exercises you learn in training your employer may provide [Davidhizar 1999].
- Develop supportive relationships with coworkers and others outside of your work environment [Davidhizar 1999].

Stress management techniques really can lower your stress level. For example, nurses trained in relaxation techniques reported a significant increase in their ability to cope with stress at work [Murphy 1983].

5.4 Resources

NIOSH. Stress topic page [www.cdc.gov/niosh/topics/stress/].

5.5 References

BLS [2008]. Occupational outlook handbook 2008–2009 Washington, DC: U.S. Department of Labor, Bureau of Labor Statistics [www.bls.gov/oco].

Brulin C, Gerdle B, Granlund B, Hoog J, Knutson A, Sundelin G [1998a]. Physical and psychosocial work-related risk factors associated with musculoskeletal symptoms among home care personnel. Scand J Carin Sci 12:104–110.

Brulin C, Goine H, Edlund C, Knutsson A [1998b]. Prevalence of long-term sick leave among female home care personnel in northern Sweden. J Occup Rehabil 8(2):103–111.

Davidhizar R [1999]. Let stress make you—not break you. Home Healthc Nurse 7(10):643–650.

El-Askari E, DeBaun B [1999]. The occupational hazards of home health care. In: Charney W, Fragala G, eds. The epidemic of health care worker injury. Boca Raton:FL CRC Press LLC, pp. 201–213.

Hurrell J, Murphy L [1992]. Psychological job stress. In: Rom W, ed. Environmental and occupational medicine. 2nd ed. Boston, MA: Little and Brown, pp 675.–674.

Jarrell RB [1997]. Home care workers: injury prevention through risk factor reduction. Occup Med: State of the Art Reviews 12(4):757–766.

Johansson JA [1995]. Psychosocial work factors, physical work load and associated musculoskeletal symptoms among home care workers. Scand J Psychol 36:113–129.

Kendra MA, Weiker A, Simon S, Grant A, Shullick D [1996]. Safety concerns affecting delivery of home health care. Public Health Nurs 13(2):83–89.

Kennedy-Malone L [1996]. The stay or stray phenomena. Home Healthc Nurse 2:103–107.

Murphy LR [1983]. A comparison of relaxation methods for reducing stress in nursing personnel. Hum Factors 25:431–440.

Neysmith SM, Aronson J [1997]. Working conditions in home care: Negotiating race and class boundaries in gendered work. Int J Health Serv 27(3):479–499.

NIOSH [1999]. Stress...at work. Cincinnati, OH: U.S. Department of Health and Human Services, Centers for Disease Control and Prevention, National Institute for Occupational Safety and Health. DHHS (NIOSH) Publication No. 99–101.

Prager SB [1996]. The vagaries of home health care: a critical review of the literature. J Long Term Home Health Care 15(1):19–29.

Sauter S, Swanson N [1996]. An ecological model of musculoskeletal disorders in office work. In: Moon S, Sauter S, eds. Psychosocial factors and musculoskeletal disorders in office work. New York: Taylor & Francis, pp. 3–21.

Stonerock C [1997]. Home health aides: home care's "endangered species." Home Care Provid 2(1):15–17.

6

Violence

6.1 Introduction

Serving patients in the community is the essence of home healthcare. Yet, the community setting makes home healthcare providers more vulnerable to violent assaults. Home healthcare workers face an unprotected and unpredictable environment each time they enter the patient's community and home. According to estimates of the Survey of Occupational Injuries and Illnesses (SOII) [BLS 2007a], 330 nonfatal assaults on home healthcare workers occurred in 2006—a rate of 5.5 per 10,000 full-time workers, more than twice the rate for all U.S. workers.

An effective violence protection program requires the following:

- The patient and family should provide a safe environment in the home.
- The worker should be able to assess risks in the environment.
- The employer should provide information about the responsibility of the patients family.
- The employer should train the staff to assess risks and implement acceptable interventions [Sylvester and Reisener 2002].

6.2 What are some factors that increase the risk of violence to home healthcare workers?

The patient's home may be in a high-risk area for violence; there may be drug traffic or high-crime areas nearby. A patient's history of mental illness, alcoholism, drug abuse, or violence may also increase the risk. More time spent in the patient's home may result in a higher risk of violence [Kendra et al. 1996]. The employer may underestimate the risks to the workers and overestimate the support they receive [Kendra 1996; NIOSH 1999].

6.3 What does workplace violence include?

The spectrum of workplace violence ranges from verbal abuse and threats of assault (by human or animal) to homicide. Examples of violence include the following:

- **Threats:** expressions of intent to cause harm (verbal, body language, written)
- **Physical assaults:** attacks including slapping, beating, rape, homicide, and the use of weapons such as firearms, bombs, knives
- **Mugging:** an aggressive assault, usually by surprise and with intent to rob

Home healthcare workers may need to resolve violence issues without immediate help from their employers or coworkers. The patients may have complex physical, psychological, psychiatric, and social needs. The potential for alcohol and drug abuse and the presence of firearms in patient homes further endangers the worker [Fazzone 2000; McPhaul 2004; NIOSH 1996]. Family issues are more likely to increase in intensity and become out of control in the home than in the hospital setting. Chaotic family relationships, poor resources or lack thereof, poor hygiene, and presence of animals all may increase risk of violence directed at or in the vicinity of the home healthcare worker.

Verbal abuse is a form of workplace violence and is a source of workplace stress. Verbal abuse may come from the patient, family members, or people in the community. Verbal abuse may be as subtle as constantly requesting that the home healthcare worker perform duties out of the scope of her or his job (such as cleaning) or complaining about their job performance or appearance.

Home healthcare workers don't always report to their employer when they meet with violence while at work. Therefore, the true extent of violence in the home healthcare industry is unknown [Lanza and Campbell 1991]. The following are reasons why violence is often not reported:

- There is no consistent definition of violence or standardized reporting procedures.
- Workers fear accusations of incompetence, or they think their employer might assume that they were the cause of the violence.
- Workers may believe that dealing with violent behavior is part of the job.
- Workers may be embarrassed and hesitant to report violent behavior.

6.4 What are some effects of this violence?

The effects of violence can range from minor to serious physical injuries to temporary or permanent physical disability to psychological trauma. Violence can even lead to death: five home healthcare workers lost their lives in 2006 because of assaults and violent acts [BLS 2007b].

Violence may also have undesirable organizational outcomes:

- Low worker morale
- Increased job stress
- Increased worker turnover
- Reduced trust of employer and coworkers

Violence or safety concerns may adversely affect the quality of patient care. If home healthcare workers do not feel safe and limit the length of time of the visits or reduce the frequency of visits, patient assessment and education will decrease. Staff may be fearful and refuse to provide services in high crime areas. All these factors may affect patient outcomes [Kendra et al. 1996; Brillhart et al. 2004].

6.5 How can I prevent and control violence in a patient's home?

In its document *Guidelines for Preventing Workplace Violence for Healthcare and Social Service Workers*, OSHA [2004] encourages employers to establish violence prevention programs and to track their progress in reducing work-related assaults. At a minimum, a violence prevention program should create a clear policy of zero tolerance for workplace violence, verbal and nonverbal threats and related actions.

6.5.1 Recommendations for Employers

- Develop a standard definition of workplace violence.

- Create a zero tolerance policy for workplace violence.
- Ask employees to report each incident, even if they think it won't happen again or it might not be serious.
- Develop a written plan for ensuring personal safety, reporting violence, and calling the police.
- Conduct training on the workplace violence plan when the employee is hired and annually thereafter.
- Let workers know about the risks of their assignments and how to assess the safety of their work environment and its surroundings.
- Train employees to recognize verbal abuse.
- Train employees to identify different types of illegal drugs and drug paraphernalia.
- Train employees to recognize the signs and body language associated with violent assault and how to manage or prevent violent behavior, such as verbal de-escalation techniques, management of angry patients, recognizing and protecting themselves from gangs and gang behavior.
- Investigate all reports of a dangerous work environment and of violent assault.
- Analyze reports of violent assault, and use them for revising safety procedures.
- Do not place workers in assignments that compromise safety. Before initiating each home health service, consider the following steps:
  — Check with the local police station about the safety of the location.
  — Obtain consultation in the case of patients with psychiatric illnesses for an assessment of the potential for violent behavior.
  — Have a social worker evaluate the family and home situation.
  — Provide security or police support if needed [Kendra et al. 1996; Jarrell 1997].
- Keep close track of staff members' schedules.
- In the case of an unacceptable home environment, advise the patient on working with social service agencies, the local police department, or family members and neighbors to make the home less hazardous so care can continue.
- Provide cell phones to all staff on duty. Reports of surveys and focus groups indicate that home healthcare workers consider cell phones to be lifelines [NIOSH 1999].
- Consider other equipment, such as employer-supplied vehicles, emergency alarms, two-way radios, and personal bright flashlights to enhance safety [NIOSH 1999; Fazzone et al. 2000].
- Establish a no-weapons policy in patient homes.
- If such a policy is not required, request at a minimum that, before service is provided, all weapons be disabled, removed from the area where care is provided, and stored in a secure location.

6.5.2 Recommendation for Workers

- If possible, visits in high-crime areas should be scheduled during daylight hours.

- Consider working in pairs in high-crime areas.
- Always know where you are going. Have accurate directions to the house or apartment.
- Always let your employer know where you are and when to expect you to report back.
- When driving alone, have the car windows rolled up and doors locked.
- Park the car in a well-lighted area.
- Park in an area away from large trees or shrubs that a person could hide behind.
- Keep healthcare equipment, supplies, and personal belongings locked out of sight in the trunk of the vehicle.
- Before getting out of the car, check the surrounding location and activity. If you feel uneasy, do not get out of the car.
- During the visit, use basic safety precautions:
    — Be alert.
    — Evaluate each situation for possible violence.
    — Watch for signals of impending violent assault, such as verbally expressed anger and frustration, threatening gestures, signs of drugs or alcohol abuse, or the presence of weapons.
- Maintain behavior that helps to diffuse anger:
    — Present a calm, caring attitude.
    — Do not match threats.
    — Do not give orders.
    — Acknowledge the person's feelings.
- Avoid behaviors that may be interpreted as aggressive (for example, moving rapidly or getting too close, touching unnecessarily, or speaking loudly).
- If possible, keep an open pathway for exiting.
- Trust your own judgment; avoid situations that don't feel right.
- If you cannot gain control of the situation, take these steps:
    — Shorten the visit. Remove yourself from the situation.
    — If you feel threatened, leave immediately.
- Use your cell phone to call your employer or 911 for help (depending on the severity of the situation).
- Report any incident of violence to your employer.
- Notify your employer if you observe an unsecured weapon in the patient's home.
- If you notice strong chemical odors or suspect that there's a drug lab in the area, notify the local police and your employer.
- If someone approaches you looking for ephedrine or pseudoephedrine, notify the local police and your employer.
- If someone approaches you looking for needles, notify your employer.
- If you are being verbally abused, ask the abuser to stop the conversation.
    — If the abuser does not stop the conversation, leave the premises and notify your employer.

6.6 Resources

NIOSH. Violence on the job. U.S. Department of Health and Human Services, Centers for Disease Control and Prevention, National Institute for Occupational Safety and Health, DHHS (NIOSH) Publication No. 2004–100d [www.cdc.gov/niosh/docs/video.html].

NIOSH. Violence: occupational hazards in hospitals. U.S. Department of Health and Human Services, Centers for Disease Control and Prevention, National Institute for Occupational Safety and Health, DHHS (NIOSH) Publication No. 2002–101. (Available in English [www.cdc.gov/niosh/docs/2002-101/] and Spanish [www.cdc.gov/spanish/niosh/docs/2002-101sp.html].)

OSHA. U.S. Department of Labor, Workplace violence [www.osha.gov/SLTC/workplaceviolence/]. An example incident reporting form is available at [www.osha.gov/Publications/OSHA3148/osha3148.html].

6.7 References

BLS [2007a]. Table R-4. Number of nonfatal occupational injuries and illnesses involving days away from work by industry, 2006. Washington, DC: U.S. Department of Labor, Bureau of Labor Statistics [www.bls.gov/iif/oshwc/osh/case/ostb1796.txt].

BLS [2007b]. Table A-1. Fatal occupational injuries by industry and event or exposure, all United States, 2006. Washington, DC: U.S. Department of Labor, Bureau of Labor Statistics [www.bls.gov/iif/oshwc/cfoi/cftb0214.pdf].

Brillhart B, Kruse B, Heard L [2004]. Safety concerns for rehabilitation nurses in home care. Rehabilitation Nursing 29(6):227–229.

Fazzone PA, Barloon LF, McConnell SJ, Chitty JA [2000]. Personal safety, violence and home health. Public Health Nurs 17(1):43–52.

Jarrell RB [1997]. Home care workers: injury prevention through risk factor reduction. Occup Med: State of the Art Reviews 12(4):757–766.

Kendra MA [1996]. Perception of risk by home health care administrators and field workers. Public Health Nurs 13(6):386–393.

Kendra MA, Weiker A, Simon S, Grant A, Shullick D [1996]. Safety concerns affecting delivery of home health care. Public Health Nurs 13(2):83–89.

Lanza ML, Campbell D [1991]. Patient assault: A comparison study of reporting methods. J Nurs Qual Assur 5(4):60–68.

McPhaul K [2004]. Home care security. Am J Nurs 104(9):96.

NIOSH [1996]. Current intelligence bulletin 57: violence in the workplace, risk factors, and prevention strategies. Cincinnati, OH: U.S. Department of Health and Human Services, Centers for Disease Control and Prevention, National Institute for Occupational Safety and Health, DHHS (NIOSH) Publication No. 96–100.

NIOSH [1999]. The Answer Group. NIOSH: home healthcare workers. Written summary and videotapes of focus group meetings of home healthcare workers (June 13 and July 7, 1999) and Chicago, Illinois (June 28, 1999). Cincinnati, OH: U.S. Department of Health and Human Services, Centers for Disease Control and Prevention, National Institute for Occupational Safety and Health.

OSHA [2004]. Guidelines for preventing workplace violence for healthcare and social service workers. Washington, DC: U.S. Department of Labor, Occupational Safety and Health Administration Pub No. 3148-01R.

Sylvester B, Reisener L [2002]. Scared to go to work: a home care performance improvement initiative. J Nurs Care Quality 17(1):75–87.

Other Hazards

7.1 Introduction

Other safety and health hazards to home healthcare workers include infectious diseases; animals; temperature extremes; poor hygiene in the patient's home; lack of running water, heat, or electricity; fall hazards; severe weather; chemical spills or acts of terrorism; and transportation hazards from daily automobile use.

7.2 What can I do to prevent and control the occurrence of or exposure to these hazards?

7.2.1 Infectious Diseases

Home healthcare workers may be exposed to infectious diseases during home visits and may even be a source of infection to the patient if the worker has an infectious disease or uses dirty equipment. Although the bloodborne pathogens standard (as discussed in Chapter 4) includes protection from blood and other potentially infectious materials, an additional infection-control-and-prevention program needs to be in place to protect home healthcare workers and their patients.

7.2.1.1 Recommendations for employers

- Implement an infection-control-and-prevention program [CDC 2007]:
    — Appoint an infection-control nurse or manager to oversee the program.
    — Provide infection-control training for healthcare staff upon hire and annually thereafter.
    — Include training on standard precautions (formerly known as universal precautions), an infection control principle that treats all blood and other potentially infectious materials as infectious.
    — Provide information about hand hygiene and cough etiquette.
    — Give training and means for healthcare staff to disinfect or sterilize their medical equipment.
- Provide all necessary personal protective equipment (for example, gloves, eye protection, masks, and respiratory protection).
- If a patient has a known case of a disease that can be spread through the air (such as TB), implement appropriate infection-control and respiratory-protection plans for the patient and worker including the following [Wurtz et al. 1996; CDC 2005]:
    — Train the worker on ways to increase ventilation in the immediate area (for example, open windows in the patient's room).
    — Inform staff about the use of proper respiratory protection (following the OSHA 29 CFR[*] 1910.134 Respiratory Protection Standard).

---
[*]*Code of Federal Regulations.* See CFR in References.

- In the event of a pandemic, such as pan-flu, reinforce your infection-control plan and enact a pandemic influenza plan. See www.flu.gov for a Home Healthcare Services Pandemic Influenza Planning Checklist.
- Consult the Centers for Disease Control and Prevention (CDC), OSHA, and the state and local health departments to prepare the pandemic influenza plan.

7.2.1.2 Recommendations for workers

- Follow your employer's infection-control plan.
- Use appropriate personal protective equipment, including medical exam gloves and respiratory protection, when necessary.
- Train patients, family members and home visitors on proper cough etiquette, hand hygiene, and social distancing.

7.2.2 Animals

In focus groups, several workers were concerned about being bitten or otherwise injured by unrestrained animals [NIOSH 1999]. Brillhart et al. [2004] reported a home healthcare worker found a snake wrapped around an I.V. pole.

7.2.2.1 Recommendation for employers

- Make restraint of animals a condition of giving home healthcare.

7.2.2.2 Recommendations for workers

- Wait outside until the pet is restrained.
- If you see fleas or other pests, discuss appropriate control measures with the patient and contact your supervisor.
- If the patient isn't receptive to pest control measures, ask your employer to contact social services to help the patient and make it possible to work there.

7.2.3 Home temperature

- The home healthcare worker may discover temperature extremes in the homes.

7.2.3.1 Recommendation for employers

- Train employees about acceptable temperature ranges and what they should do if the home they visit is extremely cold or warm.

7.2.3.2 Recommendations for workers

- If you're concerned about the home being too cold and you cannot change the thermostat, ask your employer to contact social service agencies to help the patient. Local resources may be available to help pay heating bills.
- If a home is uncomfortably warm, open the windows, use fans, and if necessary, apply cool compresses. Drink plenty of water. If you believe the patient is at risk from the heat, ask your employer to contact social service agencies to help the patient.

7.2.4 Hygiene

Hygiene may be a concern of home healthcare workers. Unsanitary homes may harbor pests including rodents, lice, scabies, or termites.

7.2.4.1 Recommendation for employers

Train employees about proper home hygiene and what they should do if the home they visit is unsanitary.

7.2.4.2 Recommendations for workers

- If a home is unsanitary, consider using clean pads with plastic on one side to set down under equipment and supplies [Brillhart et al. 2004].
- Take in only the necessary equipment and supplies so potential pests infest fewer things.
- Avoid setting things such as purses and bags on a carpeted floor.
- Use non-latex disposable gloves and hand sanitizer.

7.2.5 Lack of Water

Home healthcare workers may encounter a home with no running water or water that is of poor quality. Homes may use bottled water for drinking and have access to cisterns for flushing and bathing.

7.2.5.1 Recommendation for employers

- Train employees about potable and nonpotable water and how to ask the patient about available drinking water in their home.

7.2.5.2 Recommendations for workers

- If conditions present a health hazard, ask your employer to contact social service agencies to help the patient.
- Consider bringing several gallons of water if it is needed for patient care.
- Use hand sanitizer and do not use the toilet in a patient's home with minimal water [Brillhart et al. 2004].

7.2.6 Falls

Home healthcare workers do not have control over the walkways and may encounter icy pavement, wet floors, or wet carpeting. Loose area rugs and other floor coverings can also be hazardous for workers and for patients. The rate of lost-work days from injuries caused by floors, walkways, or ground surfaces for home healthcare workers in 2007 was 39.9, per 10,000 workers [BLS 2008a].

7.2.6.1 Recommendation for employers

- Train workers about fall protection and steps they can take to identify and reduce fall hazards for both themselves and the patient.

7.2.6.2 Recommendations for workers

- Wear sturdy, flat shoes with good slip protection.
- Walk slowly on icy or wet surfaces.
- Examine the patient's walking path to the bathrooms, eating areas, and sitting areas:
    — Remove or securely tape down rugs using double-sided tape if the patient gives you permission to do so.
    — Secure cords and any other loose materials in the walking path that could cause the patient or you to slip, trip, or stumble [Parsons et al. 2006].
- Use handrails.
- Turn on outside lights before returning to your car in the dark.
- Clean up spills as soon as they happen.

7.2.7 Severe Weather

Home healthcare workers may be exposed to severe weather including tornados, hurricanes, earthquakes, blizzards, or ice storms.

7.2.7.1 Recommendations for employers

- Create a severe weather program and train employees.

Employee training should include what to do while driving or while in a patient's home during each type of severe weather event.

7.2.7.2 Recommendations for workers

The Federal Emergency Management Agency (FEMA) recommends the following protective measures for various types of severe weather:

Tornado

- Seek shelter immediately if the area you are in is under a tornado warning.
- Go to a designated shelter area such as a safe room, basement, storm cellar, or the lowest building level.
- If there is no basement, go to the center of an interior room on the lowest level (closet, interior hallway) away from corners, windows, doors, and outside walls.
- Put as many walls as possible between you and the outside.
- Get under a sturdy table and use your arms to protect your head and neck.
- Do not open windows.
- If you are in a mobile home, leave. Mobile homes, even if tied down, offer little protection from tornados.
- If you are in a vehicle, get out immediately and go to the lowest floor of a sturdy, nearby building or a storm shelter.

Hurricanes

- Follow local evacuation orders.
- If you are in a mobile home, leave. Mobile homes, even if tied down, offer little protection from hurricane winds.

Earthquake

- Be aware that some earthquakes are actually foreshocks and a larger earthquake might later occur.
- If you are indoors:
  — Drop to the ground.
  — Take cover by getting under a sturdy table or other piece of furniture.
  — Hold on until the shaking stops.
  — Cover your face and head with your arms and crouch in an inside corner of the building if you are not near a table or desk.
  — Stay away from glass, windows, outside doors and walls, and anything that could fall, such as lighting fixtures or furniture.
- If you are outdoors:
  — Stay there.
  — Move away from buildings, streetlights, and utility wires.
- If you are in a moving vehicle:
  — Stop as quickly as safety permits and stay in the vehicle. Avoid stopping near or under buildings, trees, overpasses, and utility wires.
  — Proceed cautiously once the earthquake has stopped. Avoid roads,

bridges, or ramps that might have been damaged or destroyed by the earthquake.

Blizzard or Ice Storm

- Drive only if absolutely necessary. If you must drive, do the following:
  — Travel during daylight hours, don't travel alone, and keep others informed of your schedule.
  — Stay on main roads; avoid back-road shortcuts.
  — Use snow tires or chains when appropriate.
- If a blizzard or ice storm traps you in the car, do the following:
  — Turn on hazard lights and hang a distress flag from the radio antenna or window.
  — Remain in your vehicle where rescuers are most likely to find you.
  — Do not set out on foot unless you can see a building close by where you know you can take shelter.

7.2.8 Chemical Spills and Acts of Terrorism

Home healthcare workers may find themselves in a neighborhood that has been affected by a chemical spill or an act of terrorism. The following protective measures are recommended by FEMA in the event of a chemical or hazardous material emergency, or acts of terrorism:

7.2.8.1 Recommendations for employers
- Create a program for response to community emergencies and train employees.

7.2.8.2 Recommendations for workers
- If you are asked to evacuate an area, do so immediately.
- Stay tuned to a radio or television for information on evacuation routes, temporary shelters, and procedures.
- Follow the routes recommended by the authorities—shortcuts may not be safe. Leave at once.
- If you are told to seek shelter and you are in a vehicle, stop and seek shelter in a building.
- If you must remain in your car, keep car windows and vents closed, and shut off the air conditioner or heater.
- If you are requested to remain indoors, do the following:
  — Close and lock all exterior doors and windows.
  — Close vents, fireplace dampers, and as many interior doors as possible.
  — Turn off air conditioners and ventilation systems.
  — Stay in a room that is above ground and has the fewest openings to the outside.
  — Seal gaps under doorways and windows with wet towels or plastic sheeting and duct tape.

7.2.9 Automobile Travel

Driving from home to home exposes home healthcare workers to the risk of vehicular injury or death. The 2007 incidence rate for lost workdays from injuries caused by transportation incidents was more than 10 times higher for home healthcare workers than for hospital workers and more than 3 times higher than that of general industry workers at 17.8, 1.5, and 5.6 per 10,000 workers, respectively [BLS 2008b].

7.2.9.1 Recommendations for employers

- Enforce mandatory seatbelt use.
- Ensure that workers who drive for the job have valid driving licenses.
- Include fatigue management in safety programs.
- Ensure necessary worker training for driving specialized vehicles.
- Avoid requiring workers to drive irregular hours or significantly extended hours.
- Ensure that employer-owned vehicles are serviced on a regular basis.
- Consider providing vehicles that offer the highest occupant protection in the event of a crash.
- Provide maps or global positioning systems (GPS) to employees.

7.2.9.2 Recommendations for workers

- Use seatbelts.
- Don't use cell phones while driving.
- Avoid other distracting activities, such as eating, drinking, or adjusting noncritical vehicle controls, like the radio, while driving.
- Use detailed maps or a GPS.
- Have the car checked and serviced regularly.
- Keep the gas tank at least a quarter full.
- Carry an emergency car kit containing a flashlight, extra batteries, and flares.

7.3 Resources

CDC. Avian influenza (bird flu) [www.cdc.gov/flu/avian/].

DHHS. Employer preparedness checklists for pandemic and avian flu [www.flu.gov].

DHS. Disaster planning guide for home health care providers [www.dhs.gov/xprepresp/programs/gc_1221055966370.shtm].

FEMA. [www.fema.gov/hazards/types.shtm].

NHTSA. National Highway Traffic Safety Administration home page [www.nhtsa.dot.gov/].

NIOSH. Motor vehicle safety [www.cdc.gov/niosh/topics/motorvehicle/].

OSHA. Guidance for protecting employees against avian flu [www.osha.gov/dsg/guidance/avian-flu.html].

University of Illinois—outreach, community, and home care workers health and safety: Great Lakes Center for Occupational and Environmental Safety and Health, Chicago, Illinois [www.uic.edu/sph/glakes/ce/health&safety/index.htm].

7.4 References

BLS [2008a]. Table R–7 Incidence rates for nonfatal occupational injuries and illnesses involving days away from work per 10,000 workers by industry and selected sources of injury or illness, 2007. Washington, DC: U. S. Department of Labor, Bureau of Labor Statistics [www.bls.gov/iif/oshwc/osh/case/ostb1949.txt].

BLS [2008b]. Table R–8 Incidence rates for nonfatal occupational injuries and illnesses involving days away from work per 10,000 full time workers by industry and selected events or exposures leading to injury or illness, 2007. Washington, DC: U. S. Department of Labor, Bureau of Labor Statistics,[www.bls.gov/iif/oshwc/osh/case/ostb1950.txt].

Brillhart B, Kruse B, Heard L [2004]. Safety concerns for rehabilitation nurses in home care. Rehabil Nurs 29(6):227–229.

CDC (Centers for Disease Control and Prevention) [2005]. Guidelines for preventing the transmission of Mycobacterium tuberculosis in health care settings. MMWR 54(RR–17).

CDC [2007]. Guideline for Isolation Precautions: Preventing Transmission of Infectious Agents in Healthcare Settings. By Siegel JD, Rhinehart E, Jackson M, Chiarello L, the Healthcare Infection Control Practices Advisory Committee. Cincinnati, OH: U.S. Department of Health and Human Services, Centers for Disease Control and Prevention, [www.cdc.gov/ncidod/dhqp/gl_isolation.html].

CFR. Code of Federal Regulations. Washington, DC: U.S. Government Printing Office, Office of the Federal Register.

Parsons K, Galinsky T, Waters T [2006]. Suggestions for preventing musculoskeletal disorders in home health care workers Part 1. Home Healthc Nurse 24(3):159–164.

NIOSH [1999]. The Answer Group. NIOSH: home healthcare workers. Written summary and videotapes of focus group meetings of home healthcare workers (June 13 and July 7, 1999) and Chicago, Illinois (June 28, 1999). Cincinnati, OH: U.S. Department of Health and Human Services, Centers for Disease Control and Prevention, National Institute for Occupational Safety and Health.

Wurtz R, Lee C, Lama J, Kuharik J [1996]. A new class of close contacts: home health care workers and occupational exposure to tuberculosis. Home Health Care Manage Prac 8(2):23–31.

Conclusions

The Bureau of Labor Statistics has projected home healthcare work to be the fastest growing occupation through 2016. Home healthcare workers, including home healthcare aides, nurses, physical therapists, occupational therapists, speech therapists, therapy aides, social workers, and hospice care workers, face unique hazards delivering services in patients' homes and in various diverse communities. Persons other than the patient who are residing or visiting in the patient's home may be a risk to the worker. Home healthcare workers are susceptible to injuries. These may result from overexertion due to transferring patients into and out of bed or to assisting with patient walking or standing. Home healthcare workers may be exposed to bloodborne pathogens, needlesticks, infectious agents, latex, stress, violence occurring in the home or street, verbal abuse, weapons, illegal drugs, and they may encounter animals, temperature extremes, unsanitary conditions in the homes, lack of water, severe weather, or a response to a chemical spill or act of terrorism. The large amount of driving from home to home exposes the home healthcare worker to risks of vehicular injury or fatality.

Although the chapters in this guidance book outline specific recommendations for employers and workers to improve their safety, it is important to note that the foundation of any good safety program is a strong management commitment to the safety program. A safety committee should be formed and members should represent the cross-section of employees. Employees should have a means of discussing their safety concerns and management should have a means of providing information on the company safety plans and policies. Safety training on all the topics in this guidance book should be part of initial and on-going annual training.

A summary checklist for use by the employer and worker is provided in Section 8.1.

8 • Conclusions

8.1 Checklists For Home Healthcare Workers' Safety

Employer	YES	NO
Is there an active safety program with a safety manager and a safety committee that includes employees from across the company?	☐	☐
Does initial and annual training include safety hazards and prevention?	☐	☐
Does annual training review new safety issues identified throughout the previous year?	☐	☐
Do workers have a way to obtain necessary ergonomic equipment for the home they work in?	☐	☐
Does initial and annual training include information on latex allergies?	☐	☐
Are nonlatex gloves available?	☐	☐
Is a bloodborne pathogens plan available?	☐	☐
Is the bloodborne pathogens plan updated annually?	☐	☐
Is the bloodborne pathogens plan part of initial training?	☐	☐
Is the bloodborne pathogens plan part of annual training?	☐	☐
Are workers part of the selection process for needle devices with safety features?	☐	☐
Are workers taught how to identify stressors?	☐	☐
Are workers taught techniques to reduce stress?	☐	☐
Do workers have access to an employee assistance plan or other means of counseling support?	☐	☐
Is there a no-weapons policy for patient homes?	☐	☐
If there is not a policy prohibiting weapons in the home, is there a policy requiring weapons to be disabled and locked up before the worker arrives?	☐	☐
Is the location of a new patient researched to determine local crime statistics?	☐	☐
Are workers taught how to recognize violent or aggressive behavior and how to diffuse an angry patient?	☐	☐
Are workers taught to recognize illegal drug activities?	☐	☐
Are workers taught what to do if they feel uncomfortable about a patient's community or if they believe that they are in danger?	☐	☐
Are workers taught how to identify verbal abuse and what to do about it?	☐	☐
Has an infection control and prevention plan been developed?	☐	☐

(Continued)

Employer (Continued)	YES	NO
Has a pandemic influenza plan been developed?	☐	☐
Is there an animal-control policy requiring animals to be restrained?	☐	☐
Are workers taught how to deal with threatening weather?	☐	☐
Are workers taught what to do in the event of a chemical spill or an act of terrorism?	☐	☐
Are workers taught safe driving skills?	☐	☐
Do workers have to report all incidents and traffic offenses?	☐	☐
Has the agency verified safe driving records for all home healthcare providers?	☐	☐
Are workers' driver licenses verified annually?	☐	☐

Workers	YES	NO
Does your initial and annual training include information on the following?*		
Preventing musculoskeletal disorders	☐	☐
Obtaining ergonomic equipment	☐	☐
Learning about latex allergies	☐	☐
Reviewing the bloodborne pathogens plan	☐	☐
Promoting infection control	☐	☐
Identifying stressors	☐	☐
Reducing stress	☐	☐
Recognizing violent or aggressive behavior	☐	☐
Calming an angry patient	☐	☐
Recognizing illegal drug activities		
Knowing what to do if you feel uncomfortable about a patient's community	☐	☐

Workers (Continued)	YES	NO
Knowing what to do if you believe you are in danger	☐	☐
Identifying verbal abuse	☐	☐
Knowing what to do if you believe you are being verbally abused	☐	☐
Knowing what to do if you encounter an unsanitary home	☐	☐
Preventing slips and falls	☐	☐
Dealing with threatening weather	☐	☐
Knowing what to do in the event of a chemical spill or an act of terrorism	☐	☐
Knowing how to drive safely	☐	☐
Do you know how to report your safety concerns?	☐	☐
Do you know what to do if you are injured on the job?	☐	☐
Are sufficient patient-related ergonomic assistive devices provided?	☐	☐
Do you have appropriate personal protective equipment, including gloves?	☐	☐
Are nonlatex gloves available from your employer?	☐	☐
Do you know the symptoms of latex allergy?	☐	☐
Do you consistently follow standard precautions with all blood and potentially infectious materials?	☐	☐
Do you have a properly labeled, leak-proof, puncture-resistant sharps container?	☐	☐
Do you know what to do if you feel threatened or verbally abused?	☐	☐
Are weapons removed from the area of service (for example, bedroom, living room)?	☐	☐
Do you have a cell phone or two way radio?	☐	☐
Do you follow infection control and prevention measures (for example, hand washing)?	☐	☐
Are animals restrained in the home before you render service?	☐	☐
Do you know what to do if you find unsanitary conditions (for example, lack of heating, lack of cooling, lack of potable water, insects)?	☐	☐

Workers	YES	NO
Do you wear sturdy, low heeled, slip-resistant shoes?	☐	☐
Do you have an accurate map or global positioning system (GPS) to locate the home?	☐	☐
Do you observe your surroundings and park in well lit areas, away from visual obstructions (for example, large bushes someone could hide behind)?	☐	☐
Is your car serviced regularly?	☐	☐
Do you wear your seatbelt?	☐	☐
Do you avoid talking on a cell phone while driving?	☐	☐

---

*This suggested training list is not meant to be a substitute for regulatory training requirements.

NOTES

NOTES

NOTES

www.ingramcontent.com/pod-product-compliance
Lightning Source LLC
Chambersburg PA
CBHW081855170526
45167CB00007B/3020